EFFECTS OF NOISE ON WILDLIFE

ACADEMIC PRESS RAPID MANUSCRIPT REPRODUCTION

EFFECTS OF NOISE
ON WILDLIFE

EDITED BY

JOHN L. FLETCHER

University of Tennessee
Memphis, Tennessee

R. G. BUSNEL

National Institute for Agricultural Research
Paris

ACADEMIC PRESS   New York   San Francisco   London   **1978**
A Subsidiary of Harcourt Brace Jovanovich, Publishers

ACADEMIC PRESS, INC.
111 Fifth Avenue, New York, New York 10003

United Kingdom Edition published by
ACADEMIC PRESS, INC. (LONDON) LTD.
24/28 Oval Road, London NW1 7DX

LIBRARY OF CONGRESS CATALOG CARD NUMBER: 78–6065

ISBN 0–12–260550–0

PRINTED IN THE UNITED STATES OF AMERICA

CONTENTS

LIST OF PARTICIPANTS

D. R. Ames, Department of Animal Science and Industry, Kansas State University, Manhattan, Kansas

M. C. Busnel, Physiological Acoustics Laboratory, National Institute for Agricultural Research (INRA), Jouy-en-Josas 78, France

Rene Guy Busnel, Physiological Acoustics Laboratory, National Institute for Agricultural Research (INRA), Jouy-en-Josas 78, France

Ph. Cottereau, Medical Pathology Department, National Veterinary School, Lyon, France

David H. Ellis, U.S. Department of Interior, Fish and Wildlife Service, Tucson, Arizona

John L. Fletcher, Department of Otolaryngology and Maxillofacial Surgery, University of Tennessee Center for the Health Sciences, Memphis, Tennessee

Robin T. Harrison, Forest Service, U.S. Department of Agriculture, Equipment Development Center, San Dimas, California

Raelyn Janssen, Office of Noise Abatement and Control, U.S. Environmental Protection Agency, Washington, D.C.

Jack M. Lee, Jr., Biological Studies Coordinator, Bonneville Power Administration, Portland, Oregon

Thomas E. Lynch, Tennessee Wildlife Resources Agency, Nashville, Tennessee

Arthur A. Myrberg, Jr., Rosenstiel School of Marine and Atmospheric Science, University of Miami, Coral Gables, Florida

E. A. G. Shaw, Department of Physics, National Research Council, Ottawa, Canada

PREFACE

Impetus for the symposium on which this book is based dates back to 1971 and the work of the first editor in compiling the report "Effects of Noise on Wildlife and Other Animals" for the U.S. Environmental Protection Agency. At that time several things became apparent. First, those doing work related to the effects of noise on wildlife were scattered among several disciplines. There were biologists, psychologists, agriculture and animal husbandry personnel, wildlife personnel of all kinds, engineers, and many others. As a result, no one professional meeting had any sizable number of these workers in attendance and papers resulting from their work were published in many different technical journals, many relatively obscure and not always abstracted. A second problem was that few people were particularly concerned about the possible effects of noise on wildlife or on other animals. Following publication of the EPA report to the President and Congress on effects of noise on wildlife and other animals, professional technical organizations and the various media began to show concern. The Acoustical Society of America appointed a committee to consider problems of noise and the environment, to include possible effects on wildlife and other animals. At the 8th International Congress of Acoustics, London, 1974, the International Commission on Acoustics set up a series of working groups (WG) under their Special Committee on Problems of the Environment (SCOPE). Working Group 4, on effects of noise on wildlife, was jointly chaired by Dr. R. G. Busnel, Chief, Physiological Acoustics, National Institute for Agricultural Research, Jouy-en-Josas, France, and Dr. John L. Fletcher, Professor and Director of Research, Department of Otolaryngology and Maxillofacial Surgery, University of Tennessee Center for the Health Sciences, Memphis, Tennessee. Drs. Busnel and Fletcher met at the conference anad began planning for possible field research in the area. At a later meeting, in 1974, a tentative field research protocol, outlining suggested studies to investigate possible effects of noise on wildlife, was formulated and sent to Dr. Mattei of France, then President, International Commission on Acoustics. However, no money was available to fund the project. Then, in December 1975, Busnel and Fletcher were asked by Dr. Ira Hirsh to consider organizing and chairing a joint symposium—WG4 SCOPE meeting at the 9th ICA to be held in Madrid, Spain, July 4–9,

1977. They accepted and began planning the meeting. People known or believed to be working in the area of noise effects on wildlife and other animals were contacted to see if they had research that could be presented or knew of people working in the area. From the information thus garnered those participating in the symposium were contacted and agreed to participate.

Early in the planning stage it became obvious that people from several disciplines were working in this area, many of whom would not normally attend the ICA. It was then clear that some minimal funding would be required to provide at least partial reimbursement for the cost of attending the meeting and presenting results of their work. The Office of Noise Abatement and Control, U.S. Environmental Protection Agency, was contacted and asked to fund an update of the literature concerning effects of noise on wildlife and other animals. The update would be accomplished primarily through the presentations of the participants in the symposium and by Fletcher, who would, as part of the process of organizing and conducting the symposium, review literature published since 1971, the date of the EPA report to the President and Congress. They agreed the literature should be updated and also agreed to provide funding for the symposium—literature survey.

Thanks are due to EPA for their support in the conduct of the symposium and for their interest in wildlife. We are particularly grateful for the help from Ms. Raelyn Janssen, Office of Noise Abatement and Control, EPA. It is a pleasure to acknowledge also our great debt to Dr. Edgar Shaw, President, International Congress of Acoustics. We are also grateful to Miss Jackie King, our secretary, who had to type and retype the manuscripts, and to all the contributors and participants at the symposium. All that is worthwhile and of value in this book is a result of contributions of the many people who helped put on the symposium. The mistakes and faults contained herein are in no way attributable, however, to anyone but ourselves.

JOHN L. FLETCHER

Effects Of Noise On Wildlife

SYMPOSIUM ON THE EFFECTS OF NOISE ON WILDLIFE

E.A.G. Shaw

National Research Council of Canada
Ottawa, Canada

INTRODUCTION

In June 1971, public interest in environmental quality was brought into sharp focus by the United Nations Conference on the Human Environment at Stockholm. Quite naturally much of the delegates' time was devoted to massive problems such as water pollution, the effects of deforestation, the vast increase in the use of pesticides and other chemicals and the effects of a burgeoning human population. It was the Commission of Acoustics, through its parent body the International Union of Pure and Applied Physics (IUPAP), which suggested that noise be officially recognized as a pollutant and included in the Conference Agenda.

A year earlier, in preparation for the Conference, the International Council of Scientific Unions (ICSU) which includes IUPAP set up a broadly representative Scientific Committee on Problems of the Environment (SCOPE) charged with the following responsibilities:

a)   To advance knowledge of the influence of man on his environment, as well as the effects of these alterations upon man, his health and his welfare - with particular attention to those influences and effects which are either global or shared in common by several nations.

b)   To serve as a non-governmental, interdisciplinary and international council of scientists and as a non-governmental source of advice for the benefit of governments and intergovernmental agencies with respect to environmental problems.

1

SCOPE worked closely with the Conference Secretariat before
and after the meeting and was prepared to respond to many of
the issues raised at Stockholm. At the XIV General Assembly
of ICSU in 1972, SCOPE was directed to identify those environmental issues requiring the most urgent interdisciplinary
scientific and international efforts.[1] In developing its
program SCOPE sought advice and recommendations pertaining to
noise pollution from the Commission on Acoustics.

In 1973, the Commission identifed several areas of
study which were, in its judgement, in harmony with the purposes of SCOPE and strategic to the solution of environmental
noise problems.[2] These areas were associated with the
hearing levels of human populations as a function of age in
various environments, the propagation of noise in the atmospheric boundary layer and in buildings, the effects of noise
on sleep and other human activities and the well-being of
wildlife in an increasingly noisy world. To enhance and
accelerate the generation, integration and availability of
new knowledge in these areas, the Commission recommended the
formation of several internationally-representative working
groups composed of highly qualified scientists in the various
fields. The Commission proposal with minor changes was accepted by SCOPE and five working groups were duly formed
(See Appendix). It was intended that each group would define a scientific problem in clear terms, outline a program
of research with a timetable and budget, identify the laboratories which would participate in the program and coordinate
the work over a period of years.

The scientific area which is of interest to ICA-WG4
was outlined by the Commission in the following terms:

ACOUSTIC SIGNALLING BY WILDLIFE AND INTERFERENCE BY MAN-MADE NOISE

Information about the effects of noise on wildlife is
widely scattered through the scientific literature, frequently
inconclusive and sometimes contradictory. However, as
Fletcher has recently observed, "few if any of the suggested
effects would benefit the animal or increase his chances of
survival" (3). So, it seems reasonable to ask whether man-
made noise may, in a few cases at least, pose a danger to
species which are already under serious pressure. To provide
reasonable answers to such questions is clearly no easy
matter. It is hardly surprising that few acoustical scientists have ventured into this area.

Of the many possible effects of noise on wildlife, interference with communication seems the most promising for
further study at the present time. Acoustic signals are used

by various species as aids to navigation, to maintain or
establish contact with other members of a family or larger
social group, and to convey many types of message such as
distress or danger, the presence of food and the extent of
territory.

Signal dectection theory has made such progress in the
past twenty years that it is a simple matter to delineate the
areas in which man-made noise is likely to interfere with
acoustical signalling by wildlife. Broadly speaking, one
needs to know the nature of the acoustic signals with particular reference to their acoustic frequencies, intensities and
time patterns, the spectrum levels, time patterns and terri-
torial distributions of man-made noise, and the corresponding
data for the natural background noise due to wind and other
natural phenomenia.

A careful study of this kind has recently been carried
out by Payne and Webb (4) for the fine whale (Balaenoptera
physalus). The authors advance a series of arguments to
support the theory that members of this species use intense
20 Hz signals to communicate over distances of several thou-
sand miles under favourable conditions. Accordingly, they
suggest that all members of that species within an ocean may
form an important social unit termed a "range herd". It is
now known that ship traffic is the predominant source of
underwater noise, in the 10 - 500 Hz band, even in regions
remote from shipping, except under heavy sea conditions in
which case natural noise is predominant (5). So, it is
plausible to suppose that shipping noise has, during the past
century, considerably reduced the distances over which members
of the species Balaenoptera physalus can maintain acoustic
contact. Whether this poses a threat to the species i unknown. It would also seem reasonable to enquire whether
other species of sea mammal make use of the 10 - 500 Hz band
and, if so, whether man-made noise is a potentially important
environmental pollutant for such species.

Fletcher (3) has proposed a program of experimental research on the effects of noise on true wildlife in their
native habitat. Such a program would commence with a careful
study of wildlife species, numbers and behavioral patterns
over an extended period of time in a natural isolated wilderness area together with measurements of the background noise
levels and other appropriate environmental parameters. Noise
of the type associated with modern technology (e.g. highway
traffic noise) would then be introduced at a succession of
levels for extended periods of time during which changes in
the wildlife populations would be carefully observed.

While an experiment of this kind might be set up in al-

most any part of the world it would clearly be desirable to
undertake simultaneous experiments in a variety of types of
terrain such as (a) a northern tundra, (b) tropical forest,
(c) tropical savanna, (d) temperature deciduous forest, etc.,
(6)."

It should be noted that the proposal to form ICA-WG4
drew much of its inspiration from the document "Effects of
Noise on Wildlife and Other Animals" prepared by Memphis
State University for the U.S. Environmental Protection Agency
in 1971. A specific program of work, also inspired by that
document, was considered by the initial members of the working group when they met for the first time in July 1974 at the
8th International Congress on Acoustics in London. The program gave priority to a long-term study of the effects of
noise on animal behavior (e.g. mating, brooding, parental
care, migration and social organization). In particular it
advocated a controlled experimental field study, at a carefully selected site, extending over minimum period of three
years to be conducted by a highly qualified and properly
equipped scientific team. The estimated cost was approximately U.S. $1.7 M.

Unfortunately it soon became clear that the financial
support for working group meetings and major projects, which
had been expected to come from international bodies, was
unlikely to be available in the foreseeable future. Despite
this serious setback, the Commission on Acoustics has encouraged the working groups to continue their studies hoping
that the members of the various groups might be able to marshall other resources.

Perhaps this Symposium on the Effects of Noise on
Wildlife, organized by ICA-WG4 as part of the 9th International Congress on Acoustics in Madrid, will prove to be
a watershed in this field. If so, the credit will be largely
due to a small group of dedicated scientists actively encouraged by the U.S. Environmental Protection Agency.

REFERENCES

1. International Council of Scientific Unions: Organiza-
    tion and Activities, October 1976 (ICSU Secretariat,
    Paris) pp. 129-133.
2. International Commission on Acoustics: "Noise and the
    Environment". SCOPE-II GA/24 (revised). July 1974.
3. Fletcher, J.L.: "Effects of Noise on Wildlife and Other
    Animals", NTID 300.5 (U.S. Environmental Protection
    Agency, Washington, D.C.). December 1971.
4. Payne, R. and Webb, D. "Orientation by Means of Long
    Range Acoustic Signalling in Baleen Whales", Annals
    of New York Academy of Sciences 188:110-141, 1971.
5. Wenz, G.M.: "Review of Underwater Acoustics Research:
    Noise", J.Acoust.Soc.Amer., 51:1010-1024, 1972.
6. Commission on Monitoring, "Global Environmental Moni-
    toring", SCOPE 1 (ICSU:SCOPE) 1971.

Effects Of Noise On Wildlife

INTRODUCTION

Rene-Guy Busnel

National Institute for Agricultural Research
Jouy-en-Josas, 78, France

The protection of wildlife is a serious business and
the acoustic problems relating to human noise pollution
form but a fraction of the whole subject which is neither
a simple nor an easy one.

Before we discuss a number of theoretical approaches
and aspects of an international policy, it is necessary to
look at some aspects of animal behaviour associated with
noise. We shall give some examples to illustrate these
ideas, a few drawn from the scientific literature of the
subject, some mainly anecdotal and even trivial; but most
of them are capable of being verified.

Because this book is aimed at a large number of
readers in adminstrative services and agencies responsible
for nature protection it will be necessary to examine some
popular and relatively well-known aspects of animal behaviour in relation to noise, whose complexity makes a
scientific approach essential if it is to be intelligible.

MAN AND WILDLIFE NOISE DISTURBANCE

The first night I spent in the United States some
thirty years ago in Allerton Park, University of Illinois,
Urbana, I was disturbed by an extraordinarily loud complex
noise and I found it impossible to sleep without ear-plugs.
This noise was made by a mixed chorus of amphibians, toads,
frogs and tree frogs inhabiting a large pond close to my
bedroom. It was their breeding season. Some of the calls
reached 110 dB SPL, so loud that they approached the
threshold of pain. I was tempted to recall that period of
the middle-ages when, in Europe, the Lord of the Manor used

7

to make his serfs beat the ditches to stop the frogs calling, so that he could enjoy an undisturbed night's sleep.

Another occasion I recall was a night in a California coast hotel, at Monterrey, in the town itself, where I arrived by car at dusk. Before going to sleep I was anxious to listen to and identify a curious and unusually loud noise resembling that of a hundred dogs barking. Unable to sleep and since it was very early in the morning, I walked towards the source of the noise to arrive eventually at a pier covered with hundreds of sea lions. Most people know how the buzz of just one mosquito flying about a darkened bedroom can disturb a would-be sleeper. In town or country people may lose their early morning sleep, at daybreak, because of the acoustic activity of cockerels, passerine-birds or pigeons, and at night by that of starling colonies; in Africa by social birds that like to make their nests in villages. Each one of us has, at least once in his life, heard the noise of a mouse running about in the ceiling of a barn and found it exceedingly disturbing.

It would be pointless to multiply these examples; the acoustic activities of wildlife can at times be looked upon as noise, that is to say, as far as man is concerned, a form of pollution. They apparently do not cause damage to the ear but nevertheless constitute a nuisance that cannot be ignored.

The above considerations make it intelligible that, conversely, animal life can be profoundly disturbed by noise of human origin. Before we tackle the subject it may not be superfluous to recall that the effects of noise vary widely from one specie to another, and that even such species as are perfectly adapted to human noise can vary in their reactions.

MAJOR WILDLIFE SPECIES ADAPTED TO HUMAN NOISE

Although it is difficult specifically to separate noise caused by man from that created by man's presence, it is known that a large number of animal species are adapted to man's presence and his noise. Laboratory experiments with captive animals in restricted areas show that noise can produce a variety of physiological damage, in the first place acting on the auditory and central nervous system, later inducing stress-symptoms like the classical one originally described by Selye using various physical agents. It

is not surprising that similar results should be obtained
when noise is the stimulus, eliciting a mainly endocrinal
response. However we know little of the effects of noise
on the same species living free in nature, where animals
are in their individual umwelt, but it should be noted that
a number of observations show that, on the contrary, many
species, including a few large ones, are well adapted to
human activities and noises.

These species, like parasites, take advantage of
human activities; some of them live apparently without
damage with extremely loud human noises: rats who live in
subways, mice in industrial milling plants, crows, pigeons,
starlings, gulls, who reside on airfields, are all excellent
examples which demonstrate that these species prefer, or at
least do not avoid, a noisy environment. Many detailed reports on the wildlife population of airfields have been
published; they concern not only passerine birds and rodents but also large mammals, such as deer and a number of
large birds including raptors and vultures. (4 bis).

Even in the case of the Concord airplane birds and
rabbits on an airfield seem to pay no attention to the
noisy take-offs and landings. Many years ago Thiessen and
Shaw (21,22) reported failure when they attempted to repel
ducks from a Canadian airport by using a very loud siren.
Airfields are specific flat ecological niches free from
human presence; they constitute very attractive areas for
these species that are apparently not disturbed by the
visual stimulus of airplanes or helicopters flying above
runways or by the high levels of environmental noise.

Reports (17,18,19) concerning arctic wildlife which
describe caribou walking or running away from both fixed-
and rotating-wing aircraft, the stampeding of sheep, the
fast trotting, scattering and the panic of wolves in the
presence of a helicopter, the flushing of snow-geese and the
decrease of egg production of bald-eagles are in contra-
diction with others. The contrast between these observations and what can be seen daily on any airfield may be
explained by what appears to be a learning process in the
case of certain animal populations. Work done on arctic
wildlife is interesting enough to suggest that one and the
same population should be observed for a relatively extended period of time and on the same spot. The fact is
that the unusual noise in combination with the visual
stimulus over the heads of the animals is enough to disturb any animal, including man, and make him panic; that
special circumstance considerably diminishes our interest
in this problem. In other words, an abrupt intrusion of

any kind, even acoustic, in the daily life of most vertebrate species can produce a panic reaction; but what is
due to the noise alone is not specifically known. Taking
sound alone we have the example of the sonic boom, which
is a relatively pure acoustic stimulus, compared to the
noise from visible aircraft or helicopters.

All the published reports of well controlled experiments on the sonic boom problem make it plain that the
behaviour of domestic animals and some very shy wild species
like turkeys was unaffected by repeated sonic booms (23).
The only reported case of sonic boom effect on animals concerns honey-bees in very special circumstances. Some
French police reports contain claims of beekeepers who observed the death of beehive colonies in winter. The only
possible explanation, from a biological point of view is
that the swarm was shaken by the sonic boom and scattered
inside the hive, and that the individual bees, affected by
the very low temperature (below 10°C) were unable to re-
swarm and died because it was freezing.

In many factories numerous passerine birds (sparrows,
tits, blue tits, and others) nest under the roof where
they are sujected to a continuous 115 dB SPL noise from
engines working non-stop. Farmers using carbide guns fired
at 10 or 20 minute intervals to scare away crows and
starlings found that after a couple of days crows could
be seen perching on the gun and the birds no longer reacting to the shots.

In India, cows and monkeys living in densely populated
cities, are also parasitic to man and are not in any way
disturbed by noise of human origin. Even in the Venetian
lagoon where motor-boats are used like cars in a modern
city, fisherman catch plenty of fish. Are these fish deaf?
While this is not known it is clear that they do not avoid
the waters of Venice where, if turbidity is anything to go
by, there is water as well as noise pollution.

It is also reported that during the last century, before the bison herds were virtually eliminated, a hunter
could shoot many animals without fear of herd-reaction to
the noise of the shots. And it is well known that rats
populate cargo-boats, where they are subjected to a loud
and permanent noisy environment. Again, no information has
ever been published concerning disturbance to the migrating
grey-whale along the southern Calfornia coast where marine
traffic noise is relatively heavy.

NOISE LEARNING BY ANIMAL SPECIES

From a basic biological point of view, it is important
to mention the learning ability of many animal species and
the extraordinary ability of some to recognize and distinguish between different noises, as for example, dogs
who know the sound of the engine of their master's car.
While the animal's first reaction to a new noise source ap-
pearing in a new ecological niche is fear and avoidance, if
his other sensory systems (optical, chemical) are not
stimulated, the major vertebrates quickly learn to ignore
the noise source. However, especially in the case of noise
pollution, it is rare for a sound-source to exist in the
absence of man, and in such a case the association provokes
avoidance on the part of the animal, especially of course if
it is chased. What was said about airfields and the contrast observed in the arctic with helicopters and wing-
aircraft, can be explained by this learning faculty.

A few examples may reinforce this idea: it is well
known that hords of crows and gulls will follow a tractor
to pick worms and insect larvae, ignoring the noise of the
tractor; but they leave immediately if the driver stops his
engine and moves away. It is also generally agreed that
porpoises can learn fairly quickly the "signature" of a
boat's engine if an attempt is made to catch some members
of their school. In Europe, at least in France, the
medieval technique of deer-hunting still in use includes
loud horn noises, human voices and the barking of dogs to
chase a particular individual. The others do not move very
far away after they have located the quarry.

HUMAN NOISES ATTRACTING WILDLIFE

While there are examples of the avoidance by wildlife of noises of human origin, the opposite is by no means
rare and it is not always easy to understand the relationship existing between noise-level and frequency-pattern on
the one hand and the animal's reaction on the other, and
the correlation with their own natural acoustic signalling
system. Mosquitoes may perhaps not be thought of as a
wildlife species in our sense, but it has been reported
that engine noise attracted swarms of mosquitoes (12).
The sound of a mechanical piano and that of a motor-pump
were also reported to have attracted swarms of mole-
crickets. In those two cases we know specifically it hap-
pened; the noise spectra contained a frequency modulation
similar to the mating signals of the female (7).

Some human noises attract animals because they excite their curiosity. In Norway and Sweden steam-engine noises attracted elk and created problems for the railroad companies, because the animals strayed on the line.

As far as Hollywood movies can be trusted, the same was true of the bison and the reason why American railway engines are fitted with cow-catchers is that the animals were not frightened of train bells.

In experiments on crows and gulls, distress calls produce a reaction and when sounds were emitted from a car in the country-side, cows (and sometimes horses) were often attracted to the sound-source; the herds arrived trotting and running to the fence nearest the source, which was invariably a loud noise (about 115 dB SPL at the source). This reaction occured so often that in many cases it was found necessary to stop the emissions because the animals became so excited that the farmers complained (4).

At sea, it is well known that some porpoise species are attracted by ships and swim around the bows; they do the same with sailing-boats which are almost completely silent, but in the case of motor-boats they are attracted from a distance. They have learned to connect the propeller with its noise because they apparently like to be pushed by or swim on the front wave. They seem to be undisturbed by the propeller noise, even that of a war-ship sailing at 38 knots and consequently very noisy. But the loudness of the signal does not interfere with the learning process. It is known also that these species, although they possess a wide hearing range, are apparently not disturbed by the intense propeller noise. Some shark species have also been known to be attracted by ships, but they follow them for food. Again in this case their learning process seems to be based on an association with noise, and they do not avoid the ship.

A FEW SPECIFIC REACTIONS TO NOISE

Transient loud noises generally induce a reaction of alarm in many species, and transient vibrations like those that precede an earthquake also produce a reaction of fear. Fire-work or rocket noises, for example, created a strong fear in a number of species and caged lemurians can develop a sudden attack of diarrhea. But again some species can be trained to ignore explosive noises. That is the case of army horses, and a similar learning process can be expected in many forms of wildlife if the stimuli are repeated often enough.

Some noises produce specific reactions in some
species. The howling of domestic dogs, when they hear a
siren is well know, and this reaction is classical in
German and Belgian sheepdogs, as well as a few other
strains. But reactions to noise can vary widely according
to the biological state of a specie. Hares and partridges
generally take to flight, but in early June, when partridges are hatching and hares having their young these
species are chiefly located in lucern fields and farmers
are making hay, using heavy machinery that runs at high
speed and very noisily, producing something of the order
of 120 dB SPL near the cut-bar. At this biological stage,
both species have a fear reaction and flatten themselves
on the ground; they do not fly or run away and they get
killed by the machine. This means that the engine noise,
however loud, does not repel them because, like a similar
one at another stage, the fear reaction, at this biological
stage, is not one of flight.

That is one of the main reasons why in the study of
wildlife reaction to noise, one must take into consideration the different biological aspects of each animal
species, and the impossibility of making general statements,
or of pursuing the same research program for each one.

WILDLIFE IN A NATURALLY NOISY ECOLOGICAL NICHE

There are two cases of naturally noisy ecological
niches. The first one is given by the natural environment,
the second one is that of large animal colonies of the same
specie.

One case is the back-ground noise of the sea, mostly
a white noise generated by barnacles in shallow waters
when the bottom is rocky; it has been found sometime at
a level of around 80 dB, with a circadian-rythym. This
permanent back-ground noise does not seem to disturb any
specie, invertebrate or vertebrate (1).

Near large water-falls, which produce a continual
loud noise, there are fish and birds; are they deaf?
Nothing is known about their audiograms but it seems difficult to suppose that they suffer from loss of hearing.
At least, if they have stayed there and multiplied, they
can obviously tolerate this noisy environment.

In the second case, social noises are often very loud for species which live in dense and large colonies, chiefly social birds: flamengoes, eiders, penguins, jackdaws, ducks, geese. They apparently suffer no hearing damage or any behavior disturbance related to the noise.

NOISE AND ANIMAL COMMUNICATION

It was suggested that noise can heavily disturb animal acoustic communication.

Masking or jamming acoustic signals is really difficult, some experiments with bats demonstrate this fact (9,10). The same thing done with lovebirds shows that the birds are able to recognize their mates' signal 13 dB below a back-ground noise made with a masking noise of other love-birds (3). In the Adelie penguins, there exist rookeries counting many thousands of birds, which congregate together, in a very dense colony subjected to the heavy noise of the arctic wind, that blows semipermanently; the male sits on the eggs and the female, returning from the sea after a long period remain able to find her mate, at night, guided only by the acoustic signal, across the colony, and even if thousands of birds are calling together, making an extremely loud noise which can be heard for miles (15).

Many years ago there appeared an interesting study (5) concerning the background noise of the Panamanean Jungle, which is very heavy both during the day and at night. There the species have apparently no problems of communications and this is probably the case in all naturally noisy ecological niches.

It seems probable that most animal species can use their ability to detect their associates', partners' or mates' signals below the level of the background noise making use of the cocktail-party effect (3), within certain unknown limits for most of the species.

BEHAVIOR OF DEAF ANIMALS

Some strains of animal species are genetically deaf; they are mostly laboratory species, mice, guinea-pigs, cats and dogs.

Is their basic behavior distorted compared to that of normal hearing strains?

A genetic sub-strain of mouse: GFF dn/dn, has a reproductive level fairly similar to sub-strain GFF +/+ which
is not deaf (Table I). In mice, each strain has its own
reproductive level, and under identical conditons (habitat,
temperature, diet) GFF +/+ has a level inferior to that
of Swiss Albinos Rb3. There is a slight difference between GFF +/+ and GFF dn/dn (20,2) in favor of the hearing
strain.

But, in fact, does the mouse actively use his hearing
system? Each specie for each sequence of behavior, uses at
least one sensory system or two; frequently one is dominant.

For example, in mice, there is a retrieving behavior
of the young pups by the mother, and it was established by
Noirot, that the mother is guided to the pups by the
ultrasonic cries made by the isolated pup, which is deaf
until it is 12 days old (24).

Studying this problem, Busnel and Lehmann (2) were
able to demonstrate that, in genetically deaf mice, the
retrieving behavior is the same as in the normal hearing
strain, with practically the same efficiency (Table II).
It is also the same thing for a blind strain of mice: this
retrieving behavior is mostly based on chemical signals
and olfaction.

In fact, mice, like many if not all species, have
different sensory systems which act simultaneously and
complementarily, and for some types of behavior, use especially one or more, by correlation between the signals
they receive and their sensory abilities. Deafness, in
this case, does not significantly affect the success of
this important behavior.

That can help to explain why some species like rats
are able to survive in subways and boats if the traffic
or motor noises are so painful for humans; most of the
social behavior of this specie is preferential based on
chemical signals and the olfactive system.

THEORETICAL APPROACH

From these different examples it can be seem that
wildlife reactions to noise exposure are not easy and
simple to define or predict. Each specie and within each
one, at different stages of life, reactions to noise can
vary considerably with the season, ecological niches,
animal population density, social activities, physical

TABLE I

Comparison of the reproductive level in normal hearing mice strains
and deaf strain (GFF dn/dn) From Busnel and Lehmann (2)

Lines		Number of Females	Number of Litters	Number of pups at day 0	Mean Number of Pups per Litter at day 0	Percent of cannibalism at day 21 (weaning)
Swiss Albino $R_b 3$		67	232	1.894	8,16± 0,35 X	4,65 X
GFF +/+	full term	22	62	483	7,79± 0,29 XX	3,81
	caesarian	76	76	571	7,51± 0,19 XX	
GFF dn/dn	full term	42	163	1.041	6,39± 0,14	2,90
	caesarian	76	76	464	6,11± 0,25	

X   Significant with respect to GFF dn/dn $p < 0,05$
XX Significant with respect to GFF dn/dn $p < 0,001$

Interline comparison of number of fetuses, number of pups at birth and rate of cannibalism
between birth and weaning.

TABLE II

Comparison of retrieval scores between two sub-lines of mice: GFF +/+ with normal hearing system and GFF dn/dn, genetically deaf. Time of the detection speed expressed in seconds. From Busnel and Lehmann(2)

	Number of Females	Detection Speed: 1st Pup	Retrieval Speed: 1st Pup	Retrieval Speed: 2nd Pup	Retrieval Speed: 3rd Pup
GFF dn/dn	27	15,15+2,14	18,54+2,65	32,84+2,51	46,72+3,64
GFF dn/dn	27	13,48+1,71	16,67+1,78	33,71+2,71	48 +2,2
t		8,6799	0,5754		0,287
p		N.S.	N.S.	N.S.	N.S.

Comparison of retrieval scores in seconds between the GFF +/+ and the GFF dn/dn sublines.

parameters of the noise; an important distinction must be
made and taken into account between permanent sources, and
intermittent and mobile sources.

Studying all wildlife species at the same time is
certainly unrealistic, because each one poses different problems; it is necessary to choose the specie best suited to
the investigation of a particular problem. In fact, technics
for animal field observations exist that use mostly bio-
telemetric methods, in order to follow animal displacements
in correlation with the noise source situation.

There is need for long term research field studies because scientific literature on this topic is both scarce and
incomplete.

POLITICAL RESEARCH PROPOSAL

Governmental agencies in charge of acoustic pollution
problems are faced with a difficult problem of choice and
decision. What is the goal, how is it to be reached?
Scientifically speaking, it is not worthwhile making only one
decision for a whole country. Decision can be relative
only to a certain ecological ground, with specific biological
and climatical characteristics, and where animals populations
are known in species and numbers. Within this inventory
one must decide which are in need of protection.

From this inventory, studies may be started, in the
laboratory and in the field. A knowledge of the acoustic
sensitivity of the specie is needed, because, in most cases,
audiograms of the majority of wildlife species for example,
are completely unknown and this is the case for almost all
the large domestic mammals; it will be necessary to obtain
not only electrophysiological data, such as the auditory
evoked potential, but also others, by behavioral technics.
Behavior patterns of the major biological activities, like
pairing, mating, nesting, all the important life functions,
have to be studied and checked to record in the case of each,
the chief impacts of sound and noise, or the effects of the
acoustic stimuli on the important life functions.

In the field, by the use of appropriate technics,
scientific studies have to be made for a long term, in many
cases, 3 or 4 years, to be able to notice any learning effect, and to know correctly the secondary effects of noise,
such as animal dispersion, adaptation, changes in migration
routes and so on. The ecological impact of noise can be
measured through the expected results, but not before, and

again, only for the specie studied. All general evaluation
will be unrealistic and scientifically invalid.

Perhaps this proposal is too idealistic; in fact the
only actually known approach to this problem is the one presented at the 92nd Meeting of the Acoustical Society of
America, in 1976, concerning Arctic Wildlife, and gas pipeline related noises.

In this report one can find the first qualification for
any project of this kind: a geographic limitation to a
certain area, with known species populations.

It was decided to study only few species: Dall sheep,
moose, bear, Lapland longspur, bald-eagle, snow goose, wolf.
If the published results are incomplete and summarize only
short term studies, it seems that the basic lines of this
project are good and can be an approximate model for further
research (6,9,11,13,16).

PHILOSOPHICAL POINT OF VIEW

How much is too much was asked by H. von Gierke, and
this is an excellent philosophical attitude, but also a
scientific one, if one consideres acoustic pollution at least
for the wildlife problem. Wildlife is subjected to a heavy
pressure from much pollution originating in man's social and
technical life. The acoustic aspect alone seems to be a
minor one compared to the direct and massive intrusion of
man himself in wildlife territories, with all the characteristics of his own behavior, curiosity, distruction, killing.

Philosophically one can think also that man's pollution
is less important for wildlife than for man himself. Looking
from an airplane on our big cities, with their large suburbs
which look like bacteria in a Petri dish, are constantly
increasing, it is not difficult to accept the idea that even
in peace man is man's worst enemy creating his own pollution
with chemicals, bacteria, traffic and industrial noise,
rock-music and so on.

GEOPOLITIC'S OF WILDLIFE PROTECTION

In western Europe problems of wildlife are not basically
important because practically all large species are mostly
concentrated in parks and mountains. Large mammalian,
game animals, like deer and wild boar that live in forests
are apparently safe from the acoustic point of view, not
really disturbed by the traffic noise of highways. Airfields

are favorite niches for some bird species and rabbits. Most
of endangered species like european bison, chamois, brown
bear, mufflon, are protected in national parks. In some
cases, some species populate rapidly, individuals leave
protected area and are killed by people; organized hunting
has to be used to regulate these populations. Recent
catastrophic epidemic diseases are the consequence of over
population in parks. In other words, Europe is not directly
concerned in wildlife problems.

It is not the case for North America, Africa, the
Soviet Union, India, Australia, where local distribution
of human densities is not like the european one, and where
there are very large empty territories with numerous wildlife species.

It is encouraging to see how much American and Canadian
governmental agencies take care of the environment, devote
time and money to the study of problems of wildlife protection against pollutions. In this meeting, 96% of
participants are from these two countries.

It is a scheme in which only they are really interested,
and it is regretable that Asian and African nations, do
not pay enough attention to this problem.

Everywhere, wildlife is subjected to heavy pressure from
different kinds of pollution originating in social life, and
it is hoped that nations in charge of a wildlife natural
heritage will try to understand the problem and to protect
their wildlife before human overpopulation and human pro-
pensity to destroy everything, even involuntarily, will not
be one of the major biological phenomena of the next
century.

BIBLIOGRAPHY

1. Busnel, R.G. and Giban, J.: Protection acoustique des
   cultures etautres moyens d' effarouchement des oiseaux.
   Colloque I.N.R.A., 1958, Jouy-en-Josas.
2. Busnel, R.G. and Dziedzic, A.: Rythme du bruit de fond
   de la mer à proximité des côtes, et relations avec l'
   activité acoustique des populations d'un Cirripède fixe
   immergé. Cahiers Oceanograph., 1962, n° 5, 293-322.
3. Busnel, R.G. and Lehmann, A.: Acoustic signals in mouse
   maternal behavior: retrieving and cannibalism.
   Zeitsch. f. Tierpsychol., 1977 (in press).
4. Busnel, R.G. and Mebes, H.D.: The "cocktail party"
   effect" in intraspecific communication of Agapornis
   roseicollis vieillot (Aves, Psittacidae). Life Sc.,
   1975, 17, n° 10, 1567-1570.
5. Eyring, P.: Jungle's acoustics. J.A.S.A., 1946, 18 ,
   n° 2, 257.
6. Environmental Protection Agency, 1971. Effects of Noise
   on Wildlife and Other Related Animals, NTID 300.5,
   December 31.
7. Feytaud, J.: La courtiliere et les moyens de la combat-
   tre. Rev. Zool. Agri., 1933, 5.
8. Gollup, M.A. and Davis, R.A. 1974, Gas Compressor Noise
   Simulator Disturbance to Snow Geese, Komakuk Beach,
   Yukon Territory, Sept. 1972, Disturbance to birds
   by Gas Compressor Noise Simulators, Aircraft and Human
   Activity in the Mackenzie Valley and the North Slope,
   1972, ed. by W.W.H. Gunn and J.A. Livingston, Arctic
   Gas Biological Reprot Series, Vol. 14, Canadian Arctic
   Gas Study, Ltd. and Alaskan Arctic Gas Study Co.
9. Griffin, D.R.. McCue, J.J.G., and Grinnell, A.D.: The
   resistance of bats to jamming. J. Exp. Zool., 1963, 152,
   n° 3, 229-250.
10. Griffin, D.R.: Learning in the dark. Yale Univ. Press,
    1958.
11. Lenarz, M., 1974, The reaction of Dall sheep to an FH-
    1100 Helicopter, The Reaction of Some Mammals to Air-
    craft and Compressor Station Noise Disturbance, ed.
    by R.D. Jakimchuk, Arctic Gas Biological Report Series,
    Vol. 23, Canadian Arctic Gas Study, Ltd., and Alaskan
    Arctic Gas Study Co.
12. Maxim, H.S.: Mosquitoes called in by a dynamo noise,
    Nature, 1901, 64, 655.
13. McCourt, K.H., Feist, J., Doll, D., and Russel, J.J.:
    Disturbance Studies of Caribou and Other Mammals in the
    Yukon and Alaska, 1972, Arctic Gas Biological Report
    Series, Vol. 5, Canadian Arctic Gas Study, Ltd. and
    Alaskan Arctic Gas Study Co, 1974.

14. Noirot, E.: Ultrasounds and maternal behavior in small rodents, Developmental Psychobiol., 1972, $\underline{5}$, $n^{o}4$, 371–387.

15. Prevost, J.: L'Ecologie du manchot empereur. These Doctorat d'Universite., Paris, 1961.

16. Reynolds, P.C.: The effects of simulated compressor station sounds on Dall sheep using mineral licks on the Brooks Range, Alaska, The Reaction of Some Mammals To Aircraft and Compressor Station Noise Disturbance, ed. by R.D. Jakimchuk, Arctic Gas Biological Series, Vol. 23, 1974, Canadian Arctic Gas Study and Alaskan Arctic Gas Study Co.

17. Ruth, J.S.: Reaction of arctic wildlife to gas pipeline related noise. News Release, 1976, November 18. DD3.

18. Salter, R. and Davis, R.A.: Snow geese disturbance by aircraft on the north slope, September, 1972, Disturbance to Birds by Gas Compressor Noise Simulators, Aircraft and Human Activity in the Mackenzie Valley and the North Slope, 1972, ed. by W.W.H. Gunn and J.A. Livingston, Arctic Gas Biological Report Series, Vol. 14, 1974, Canadian Arctic Gas Study, Ltd. and Alaskan Gas Study Co.

19. Shillito, E.: Sensory involvement in Ewe-lamb recognition, Proceed of Soc. Veter. Ethol. in Br. Vet. J. 1977, $\underline{133}$, 190.

20. Strains obtained from University College, London, 1974: GFF dn/dn subline was derived from a deaf-mutant discovered in GFF line by N.S. Dedl and W. Kocher J. of Heredity, 1958, $\underline{2}$, 463–466.

21. Thiessen, G.J. and Shaw, E.A.G.: Acoustic irritation threshold of ringbilled gulls, J.A.S.A., 1957, $\underline{29}$, 1307.

22. Thiessen, G.J., Shaw, E.A.G., Harris, R.D., Gollop, J.B. and Webster, H.R.: Acoustic irritation threshold of Peking ducks and other domestic and wild fowls, J.A.S.A. 1957, $\underline{29}$, 1301.

23. Welch, B.L., and Welch A.S.: Physiological effects of noise, Plenum Press, 1970.

PHYSIOLOGICAL RESPONSES TO AUDITORY STIMULI

D. R. Ames

Department of Animal Science and Industry
Kansas State University
Manhattan, Kansas

INTRODUCTION

Data for humans and laboratory animals suggest that
the sound environment may elicit physiological responses
which could affect the well-being of other animals (7,26).
Reports characterizing responses of non-laboratory animals
to sound are limited with most studies dealing with sonic
booms. In many cases physiological responses for both
domestic food producing animals and/or wild animals can
only be predicted from data collected with laboratory
rodents which may have different auditory thresholds or
may be different responses to auditory stimuli. It was
the goal of studies presented here to characterize physiological responses to varied sound environments for large
domestic animals using sheep as an experimental species.
The results of this type research should aid in characteri-
zation of optimum sound environments for animals and predict the consequences of sound environments which are
stressful.

EXPERIMENT I - AUDITORY THESHOLDS

Since limited information is available describing the
auditory threshold for large animals, we felt it was
necessary to determine auditory thresholds for sheep to
insure that sound treatments used to provide physiological
responses were in fact being perceived by the experimental
animals. To do this auditory acuity trials on 10 Suffolk
ewes were conducted in a semi-sound proofed, thermally-
controlled room (21 C) using a free field technique. The
background noise level of the semi-sound proofed chamber

was 26 decibels (dB).

Sound frequencies were generated by an audio os-
cillator (General Radio Model 210) with the signal amplified (Rauland-Borg Company Amplifier Model SA51A, 2S-70).
The speaker (Heath Model AS-48) capability ranged from 40
to 20,000 cycles per second (Hz) and was suspended in the
center of the room two meters above the floor. Sound
intensity was determined using the C scale of a sound level
meter (General Radio Model 1551 C). The sound system
was tested for overtones and distortions which did not
occur at the intensities studied.

Auditory thresholds were determined by two methods:
(1) Changes in electoencephalograph patterns (EEG) and (2)
behavior responses. EEG's were monitored by a telemetry
system (E & M Physiograph) on a rectolinear strip chart
recorder. Needle electrodes were placed anterior the
poll and in the area of the supertemporal plane of the
cerebral cortex. Changes in EEG patterns were identified
by simultaneously integrating the area of the EEG pattern.
Changes in the integrated recording were interpreted as
perception of auditory stimuli. Behavioral responses
(viewed through a one-way window) such as pricking the
ears, turning the head, or looking at the speaker were
observed and considered to indicate auditory perception.
(Note: In more recent investigations with cattle (20) we
have found changes in respiratory patterns to be the most
definitive technique for determining hearing threshold
for large animals).

Frequencies of 100, 200, 500, 1,000, 2,000, 5,000,
6,000, 7,000, 10,000, 11,000, and 12,000 Hz were selected
for recording and observations. Order of frequencies in
a given test period were chosen at random. During a given
test period at a specific frequency, power was increased
by an amplifier until a change in the EEG integration
pattern or behavioral reaction was noted, then dB level
was recorded. Three tests at each frequency were conducted
on each animal. There was a minimum of 20 minutes between each frequency test and a second sheep was present
during each test to avoid exciting the test animal. A
total of 330 observations were made on 10 sheep.

The threshold curve (Figure 1) declined gradually
from 100 Hz to 500 Hz, then fell more rapidly and reached
its lowest point at 7,000 Hz. Threshold values increased rapidly as frequency increased above 7,000 Hz.

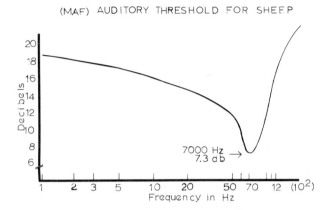

(MAF) AUDITORY THRESHOLD FOR SHEEP

Figure 1 - Plot of minimum audible field with decibels
shown as sound pressure level above background
(26 dB).

Decibel levels above background (26 dB) were highest (18.
5 dB) at 100 Hz and lowest (7.3 dB) at 7,000 Hz. There
were significant (P < .05) differences among individual
sensitivities at different frequencies with the large
variations at lower frequencies. Results from methods
of determining auditory perception (EEG pattern changes
and behavioral responses) correlated highly (r=0.95).
The value of auditory thesholds presented here is to
document the audio environment of a given animal without
wrongly assuming that man and animal interpret or respond
to the sound environment in a similar fashion.

EXPERIMENT II - HEART AND RESPIRATION RATE

Initial physiological responses to sound measured
in sheep were heart and respiratory rate. It was
reasoned that these two easily measured physiological
responses would be valuable in documenting audio stimulated response and in determining the presence (or absence) of acclimation to sound. Growing lambs were used
in trials involving three sound types: (1) United States
of America Standard Institute White Noise (USASI); (2)
Instrumental Music; and (3) Intermittent Miscellaneous
Sound (IMS) - each at two levels of intensity (75 and 100
dB). The USASI spectra was produced by a Random noise
generator (General Radio Company, Type 1381) and was

played continuously during the test periods. Instrumental
music was recorded on an 8-track, 70-minute tape and was
played continuously during exposure. The IMS noise,
which was recorded on an 8-track, 70-minute tape, included
electrical and diesel engines, jet and prop airplanes,
roller coasters, stadium noise, fog horns, fire crackers,
machine guns and cannons, rain, and band marches. The
sounds were recorded with test noise durations of 15
seconds to 3 minutes and intermittent quiet from 1 to
15 minutes. The IMS tape furnished 11 hours of noise and
13 hours of quiet during 24 hours. Lambs were subjected
to the following intensity regime: (1) 21-day control
period with background noise at 45 dB; (2) 12-day test
period at 75 dB; (3) 2-day adjustment (45 dB); (4) 2-day
test period at 100 dB. During the two-day adjustment
period before the 100 dB test, five lambs not previously
exposed to any sound treatments (nonacclimated) were
added to 100 dB treatment to compare with lambs previously
exposed to sound treatment (acclimated) to detected
acclimation to sound. Heart rate and respiration rate
were monitored before each sound treatment; 15 minutes,
1, 4 and 8 hours after treatment began; then daily during
the 12-day test. Heart rate was determined by monitoring
the EKG via telemetry. Respiratory rate was determined
by counting flank movements. Analysis of variance and new
multiple range test (13) were used to compare treatment
means.

Heart rate did not change upon initial exposure to
USASI noise for either 75 dB or 100 dB in acclimated
animals. For nonacclimated lambs 100 dB initial exposure
significantly ($P < .05$) increased heart rate to 145 beats
per minute. During the 12-day test (Table 1), heart rate
for nonacclimated lambs subjected to 100 dB intensity
sound was significantly higher ($P < .01$) than for acclimated lambs subjected to either the 75 dB or 100 dB.
Acclimated lambs' respiration rate remained constant (39
breaths per minute) when subjected to 75 dB intensity
initially, then increased rapidly after the first hour,
peaking at 64 breaths per minute by the eighth hour
(Figure 2). Nonacclimated lambs showed little change
until the fourth hour when respiratory rate increased
rapidly from 37 to 60 breaths per minute. After 8 hours
of 100 dB exposure, respiration rate for both acclimated
and nonacclimated lambs was significantly higher ($P < .01$),
than for controls and for lambs exposed to 75 dB. This
trend continued with respiration rates for the 12 days
significantly ($P < .01$) higher for 100 dB acclimated lambs.

TABLE I

Effect of sound type and intensity on heart and respiration rates.*

Levels	Types	Heart Rate	Respiration Rate
		beats/min.	breaths/min.
75 dB	USASI	121 ± 10.8^{ax}	43.3 ± 5.4^{ax}
	Music	111.7 ± 5.6^{ay}	61.0 ± 6.3^{by}
	IMS	119.0 ± 15.9^{x}	65.0 ± 20.4^{by}
100 dB	USASI	122.0 ± 10.4^{ax}	62.4 ± 15.5^{bx}
acclimated	Music	116.0 ± 8.6^{by}	44.0 ± 4.9^{a}
	IMS	123.0 ± 14.6^{x}	49.0 ± 13.3^{a}
100 dB	USASI	130.6 ± 13.2^{b}	39.0 ± 5.9^{a}
non-	Music	124.0 ± 8.3^{c}	45.0 ± 5.2^{a}
acclimated	IMS	121.0 ± 11.3	46.8 ± 8.3^{a}

* Mean and SD of 3 observations during 12-day tests.

[a,b,c] All letters within column with different superscripts are significantly different (P<.05) for intensity levels.

[x,y] All letters within columns with differnt superscripts are significantly different (P<.05) for types of sounds.

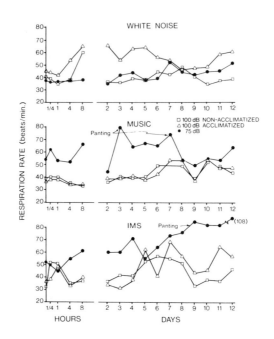

Figure 2 - Respiration rates of lambs exposed to different sound types and intensities.

Initial exposure to 75 dB music increased heart rate from 114 to 125 beats per minute, stabilizing at 122 beats per minute (Figure 3). Acclimated lambs subjected to 100 dB intensity music, compared with those subjected to 75 dB, had decreased heart rate during the first 8 hours of treatment but over the 12 days their heart rate was significantly higher (P <.05). Nonacclimated lamb's heart rate dropped from 107 to 96 beats per minute with the onset of treatment, but rose to 124 beats per minute during the 12 days and was significantly (P <.05) higher than heart rate of lambs subjected to 75 dB or 100 dB acclimated groups. Lambs subjected to 75 dB intensity sound had higher respiration rates than did 100 dB acclimated lambs, 100 dB nonacclimated lambs, or control lambs (Table 1, Figure 2).

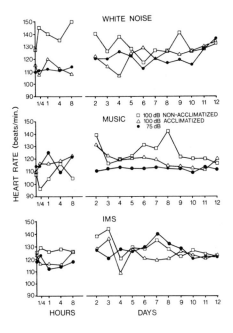

Figure 3 - Heart rates of lambs exposed to different sound types and intensities.

During the 12-day test with IMS there was no significant difference(P ⊀ .05) in heart rate between 75 and 100 dB intensities. However, heart rate remained higher for non-acclimated than for acclimated groups at 100 dB. Respiration rate was highest at 75 dB intensity, during the initial phases of the test and remained significantly higher (P ⊀ .05) throughout the 12-day trial than rates of control lambs or lambs exposed to 100 dB intensity levels.

Homeostatic responses to various environmental variables (light, temperature, pressure) are well-documented. Logically, other aspects of the external environment (i.e., sound) which are neurally sensed and interpreted by animals may also alter physiological function. Responses to sound may be considered different from variables such as temperature and barometric pressure for the latter directly involve homeostatic mechanisms in addition to being classified as "stressors"[1] under extreme situations. On the other hand, sound does not impinge directly on homeostatic mechanisms and, therefore may be viewed as a stressor that alters physiological function via adrenal responses, although evidence of sound affecting neuro-endocrine function (21,26) indicates additional avenues of action. The possibility of direct neural control of heart and respiratory rate in response to sound seems unlikely.

Data presented here indicate that at least some physiological responses (i.e., heart rate and respiration rate) to sound are commensurate with typical responses to stress. That is, evidence of adrenally oriented responses, as well as acclimation to the audio environment. For example, sudden changes that alarm or frighten usually result in either tachycardia via action of catecholamines or bradycardia due to vagal inhibition. In this study, sound exposures usually resulted in vagal inhibition except when nonacclimated lambs were subjected to 100 dB white noise. Then vagal inhibition was short-lived and increased heart rates appeared in 15 minutes. Immediate tachycardia lasted longer. Immediate heart rate response to music varied less than response to white noise, suggesting that more soothing sounds are less stressing than other types. Variation was greatest for 100 dB nonacclimated lambs regardless of sound type. Heart rate during music exposure was less variable and significantly ($P < .05$) lower than responses to white noise or IMS. Acclimation to both sound types and intensities was apparent. Regardless of type, level, or acclimation, heart rate following the 10th day of exposure varied little. These data indicate that lambs were affected by sound and that sound may be considered a stressor. In addition, it appears that there is differentiation in response to sound level, intensity, and duration.

EXPERIMENT III - GROWTH

While growth results from many integrated physiological processes, it is a necessary component of animal function and is indicative of a general compatability of animal and environment. In addition, growth rate is of major importance

------

[1] A stressor here is defined as any stimulus which provokes presponse similar to those attributable to increased levels of ACTH.

in domesticated food animals, therefore, growth trials were conducted in lambs exposed to varying sound environments.

Sixty crossbred lambs averaging 25.3 kg were used to study the effect of sound type and intensity on growth rate. The sound exposures and experimental regime were the same format as for experiment II with trials conducted in an environmentally controlled room maintained at 21C and with all animals shorn bi-weekly. Average daily gain and feed intake data were analyzed by the least square analysis of variance and Duncan's multiple range test (12).

USASI noise at 75 dB intensity significantly (P<.05) increases average daily gain and efficiency of gain compared with both controls and the 100 dB treatment (Table 2). When lambs (acclimated and nonacclimated) were subjected to 100 dB efficiency of growth was significantly lower than at 75 dB but higher than during the control period. There was no significant (P<.05) difference in voluntary feed intake among individually fed groups.

Music played continuously at either 75 dB or 100 dB (acclimated and nonacclimated lambs) had no significant (P< .05) effect on growth or efficiency of gain (Table 2). Individually fed lambs subjected to 100 dB intensities gained the least, 0.18 and 0.12 kg per day for acclimated and non-acclimated, respectively. With a high degree of variability, no significant differences in feed efficiency were observed.

Lambs subjected to IMS noise consumed less feed per day than those subjected to USASI noise and music (Table 2). In IMS exposed lambs average daily gain was significantly higher (P<.05) at 75 dB intensity, compared with controls and 100 dB treatment. Nonacclimated lambs gained an average of 0.15 kg a day; acclimated lambs, gained 0.22 kg per day.

Pooled data indicates that sound type and intensity significantly (P<.05) affected daily growth rate (Figure 4 and 5). As shown, USASI noise significantly (P<.05) increased average daily gain, compared with all other groups including controls. A sound intensity of 75 dB improved average daily gain significantly (P<.05) compared to controls and 100 dB treatments. The difference between acclimated and nonacclimated 100 dB groups was significant (P<.05).

Some reports on animal response to ambient sound indicate various physiological and psychological effects (2, 25, 26). It has been found (9) that sonic booms have no

Table 2. Effect of Type and Level of Sound on Growth Rate and Voluntary Intake of Lambs.[a]

| | | Means with S.D. (kg) | | | | Treatment levels | | |
| | | Control background noise 45 dB | | 75 dB | | Acclimated lambs 75 and 100 dB | | Non-acclimated lambs 100 dB |
Type	Initial weight	Individual	Lot-fed	Individual	Lot-fed	Individual	Lot-fed	Individual
USASI	24.8±1.1							
No. animals		5	10	5	10	5	10	5
Avg. daily gain		0.23±0.11^x	0.23±0.07^x	0.42±0.06^{by}	0.41±0.09^{by}	0.30±0.02^x	0.30±0.04^x	0.27±0.08^b
Feed intake/day		1.4^b		1.4^b		1.7^b		1.4^b
MUSIC	27.9±2.8							
No. animals		5	10	5	10	5	10	5
Avg. daily gain		0.23±0.08	0.23±0.04	0.21±0.10^c	0.24±0.13^c	0.18±0.14	0.24±0.12	0.12±0.06^c
Feed intake/day		1.7^b		1.5^b		1.5^b		1.7^c
IMS	24.5±2.8							
No. animals		5	10	5	10	5	10	5
Avg. daily gain		0.16±0.22^{xy}	0.21±0.04^x	0.39±0.13^{bz}	0.33±0.10^{by}	0.22±0.05^{xy}	0.28±0.09^x	0.15±0.22^x
Feed intake/day		1.0^{cx}		1.3^{cy}		1.1^{cy}		1.3^{by}

[a] Individual and lot-fed lambs compared separately.

[b,c] All values with different superscripts within columns are significantly different at (P<.01).

[x,y] Values within rows with different superscripts are significantly different at (P<.05).

[1] Mean of replicatial lots.

32

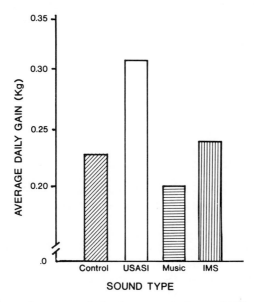

Figure 4 - Growth rate of lambs exposed to different sound
            types.

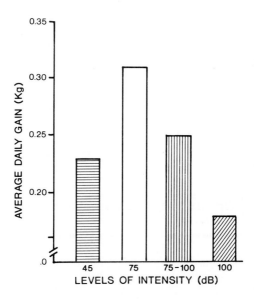

Figure 5 - Growth rate of lambs exposed to different sound
            intensities (75-100 denotes lambs acclimated to
            75 dB before exposure to 100 dB).

significant effect on feedlot animals; Sackler et al. (21)
noted that noise altered growth in rats. Data here supports
the conclusion of several workers (8, 9, 18) indicating acclimation to sound with respect to daily growth rate. The
finding that all nonacclimated animals exposed to 100 dB
gained significantly (P < .05) less than did those previously
exposed to the 75 dB treatment indicates the need for long
term experiments with sound.

It is difficult to explain altered performance of lambs
exposed to various types and intensities of sound. The
data is based on an acute response (12 days), which cannot
answer the question of long-term effects of sound. Undoubtedly, sound initially affects the nervous system and,
in particular, the auditory system. Most likely integration
of auditory inputs occurs at the midbrain level, resulting
eventually in altered endocrine function, as suggested by
several workers (1, 4, 11, 21). Another plausible avenue
of effect, neural in nature, involves efferent impulses from
higher integration centers regulating physiological function.
In this realm the digestive system itself could be directly
affected; a theory examined in experiment IV.

EXPERIMENT IV - DIGESTIVE RESPONSES

The literature contains no data on the digestibilities
of feedstuffs by animals exposed to sound stress. In humans,
a reduction in peristaltic contractions of the stomach and
gastric and salivary flow was observed when noise levels were
at or above 60 dB (14,22). Because such stimuli could affect
the digestive system, we (10) studied the effects of sound
intensities and types on nutrient digestibility coefficients,
nitrogen retention, and rumen motility in sheep. To accomplish this four yearling wethers weighing 45 to 48 kg
were used in nine trials involving three sound types. Metabolism studies were conducted in a 10 m x 12 m room with
temperature maintained at 21C. Exhaust fans run continuously
created a background noise at 45 dB. Heating and air
conditioning units did not influence background noise level.
Sound types (white noise, instrumental music, and intermittent miscellaneous sound) were as described previously.

A control metabolism trial (background noise, 45 dB)
preceded each sound-type trial at 75 or 100 dB. For the
control trial, animals were placed in metabolism crates, fed
a pelleted (.95 cm) ration and exposed to 45 dB for 14 days;
during the last 7 days, feces and urine samples were collected. Animals were then placed in individual pens and fed

rations identical (in amounts) to those consumed during the
control trial; exposure to sound was coordinated with that
of the intensity regime described previously. Wethers were
placed in metabolism crates 24 hours prior to changes in
intensity regimes; they were removed after a 7-day collection
period. This procedure was repeated for each sound type.
The sequence of study was: music, IMS, USASI.

Excreta and feed were analyzed for proximate components
(3). The energy values of feedstuffs and feces were determined in a Parr bomb calorimeter. Metabolizable energy
values were calculated from the equations of Blaxter (6)
and Paladines et al. (17). Urinary creatinine was determined
by the method of Owen et al. (16).

To measure rumen motility intraruminal pressure changes
in three fistulated wethers (60 to 65 kg), were recorded
(DMP-4A Physiograph) using a SensoTec pressure transducer
(Series 4, Model M-6BW) mounted in the rumen via fistula.
The strongest contraction of the motility cycle was used to
determine cycles/minute. Two 15-minute recordings were
taken: one an hour after the morning feeding, and the other
immediately before the evening feeding of a pelleted ration.
Following a 7-day acclimation by wethers to restraint and
ration, control readings were taken for 5 consecutive days
to establish "normal" values (background, 45 dB). During
the first treatment, USASI white noise, at 100 dB was
generated continuously for 7 days. Motility measurements
were taken twice daily for 7 days. Following a 3-week rest
sheep were exoposed to a pure tone (7000 Hz) at 100 dB.
Sound was generated for 2 minutes for every 10, and motility
recorded. Data were analyzed by least squares analysis of
variance as modified for computer by Kemp (12). Differences
among treatment means were obtained by Duncan's LSD (23).
Differences were considered significant at the 5% probability level.

Results of metabolism trials are summarized in Table 3.
Since data from the control periods (background, 45 dB) did
not differ statistically, they were pooled. Dry matter intake of sheep was less (P $<$ .05) when they were subjected
to 75 or 100 dB of each (1,083 g) compared to 45 dB (1,103 g).
Type of sound had no effect (P $<$ .10). Water intake (P $<$ .03)
and urinary output (P $<$ .02) depended on sound type; animals
subjected to IMS consistently drank more water and excreted
more urine than did controls or those subjected to the
continuous sounds, USASI and music. Level (P $<$ .02), type
(P $<$ .001), and their interaction (P $<$ .005) influenced fecal

Table 3. Dry Matter Intake, Water Metabolism, Nutrient Digestibilities and Nitrogen Balance of Sheep Exposed to Sound

Item	Control 45 dB	USASI 75 dB	USASI 100 dB	Music 75 dB	Music 100 dB	IMS 75 dB	IMS 100 dB	SE
Dry matter and water								
Dry matter intake, g	1103[a]	1053[b]	1090[b]	1095[b]	1072[b]	1101[b]	1087[b]	10.2
Water intake, ml	3127[a]	2992[a]	2302[a]	3067[a]	3149[a]	4806[b]	3504[b]	482.0
Urine volume, ml	1557[a]	1492[a]	1155[b]	1318[a]	1355[a]	2035[a]	1748[b]	198.1
Fecal water, ml	551[b]	449[c,d]	560[b]	631[a]	494[c]	393[d]	419[d]	25.1
Water excretion, ml	2108	1940	1715	1949	1849	2429	2167	208.0
Water retention, ml	1018	1052	587	1118	1300	2377	1336	384.4
Apparent digestibility coefficients								
Dry matter, %	69.52[a]	66.25[a]	66.50[a]	65.89[a]	69.27[a]	73.40[b]	69.89[b]	1.09
Gross energy, %	66.42[a]	66.98[a]	64.75[a]	63.12[a]	67.70[a]	70.18[a]	68.86[a]	
Organic matter, %	70.54[a,b]	66.80[a,b]	67.74[a,b]	66.72[a]	70.22[a]	73.69[b]	71.38[b]	1.09
Crude fiber, %	30.77[a,b]	34.94[a,b]	24.83[a,b]	28.11[a]	28.87[a]	37.51[b]	33.53[b]	2.32
Nitrogen free extract, %	83.12[a]	80.84[a]	82.42[a]	77.36[a]	82.14[a]	84.95[b]	83.56[b]	1.26
Crude protein, %	63.73[a]	61.81[a]	62.11[a]	58.04[a]	65.01[a]	67.95[b]	66.07[b]	1.35
Nitrogen balance								
Intake N, g	29.14[a]	28.11[a]	29.19[a]	27.57[a]	28.45[a]	29.42[b]	29.39[b]	.27
Fecal N, g	10.56[a]	10.74[a]	11.06[a]	11.56[a]	9.95[a]	9.44[b]	9.97[b]	.42
Absorbed N, g	18.58[a]	17.37[a]	18.13[a]	16.01[a]	18.50[a]	19.98[b]	19.42[b]	.44
Retained N, g	4.66	5.48	3.19	3.47	4.61	5.02	5.84	.84

a,b,c,d Values in each row having different superscripts are different (P<.05).

(Harbers et al., 1975)

moisture. Less fecal water was present when lambs were subjected to 100 dB music or IMS than to background noise
(45 dB), but the 100 dB USASI treatment was similar to the
45 dB. At 75 dB, more fecal moisture was excreted when the
lambs were subjected to music and less to USASI and IMS,
compared with background noise. Treatment means were not
statistically different for water excretion (fecal
moisture plus urine volume) and water retention (intake
minus excretion). Fecal water was not related to water intake. Urinary volume, water excretion, and retention (which
increased linearly with water intake) apparently are similar
to the changes in respiration rates found with nonacclimated
lambs in the same room. Pollack and Bartless (19) found that
interrupted noise or discontinuous tones are more annoying
than steady noises. If this is true for sheep, it might be
an explanation for them drinking more water.

Apparent digestibility coefficients (Table 3) revealed
a consistent trend: statistically higher coefficients were
evident when sheep were subjected to IMS sounds, compared
with the controls or with the continuous stimuli from USASI
white noise and music. Apparent crude fiber digestibility
coefficients were lower (P $<$.03) when lambs were subjected
to IMS. Neither differed from those for controls of USASI
noise. Sound intensity did not affect any apparent digestibility coefficients.

When data were analyzed using dry matter intake as a
continuous variable, water intake approached statistical
significance with sound type (P $<$.08). Apparent nutrient
digestibility coefficient analyses showed no change with
this analysis; thus, the high digestibility coefficients for
sheep subjected to intermittent sound suggest that the
digestive system is influenced by those types of auditory
stimuli. The increased digestibility of feed, and possibly
the water retention (expired moisture was not measured),
may partially explain the imporved gain of lambs when exposed to IMS but not when exposed to USASI white noise (5).

Nitrogen metabolism corroborates differences in sound
type found with crude protein digestibility: less fecal
nitrogen (P $<$.02) and more absorbed nitrogen (P $<$.01) for
sheep when exposed to IMS, even though nitrogen intake reflected changes in dry matter intake. No statistical difference was found for urinary nitrogen (P $<$.36) and nitrogen retention (P $<$.10). When nitrogen intake was held
constant during statistical analyses, no differences were
evident in fecal and absorbed nitrogen values (P $<$.08) sug-

gesting that nitrogen metabolism was not influenced appreciably by auditory stimuli.

Values for metabolizable energy and daily urinary creatinine excretion are presented in Table 4.

TABLE 4

Effect of sound type and level on metabolizable energy and urinary creatinine.[a]

Levels	Types	Metabolizable energy	Urinary creatinine-N
		kcal/day	mg/day
45 dB	Background	2651[a]	281[b]
	USASI	2567[a]	213[c]
75 dB	Music	2522[a]	456[a]
	IMS	2813[b]	177[c]
	USASI	2516[a]	171[c]
100 dB	Music	2686[a]	282[b]
	IMS	2758[b]	191[c]
SE		62	43

[a]Mean of four observations   (Harbers et al.,1975)

IMS increased metabolizable energy by 100 kcal/day over continuous sounds (P $<$.01) but intensity of sound had no effect (P $<$.90). Urinary creatinine values were highest (P $<$.01) when lambs were subjected to music: more creatinine was excreted at 75 dB than 100 dB. Creatinine values at 100 dB music were similar to those for sheep subjected to background noises. Subjecting lambs to USASI and IMS produced significantly lower values, compared with controls (P $<$.05). These values tend to confirm the growth responses observed for USASI and IMS (5), possibly indicating reduced protein catabolism. Since the sequence of sound stress was music, IMS, and USASI, possibly these data reflect acclimation by the sheep to both sound and metabolism stall stresses.

The number of rumen motility cycles within 15 minutes were determined in three sheep twice daily during 7-day trials. No differences in rumen motility were evident when sheep were subjected to either continuous or intermittent sound at a level of 100 dB (P $<$.34) (Table 5). Within each

TABLE 5

Effect of continuous and intermitten sound stress of frequency of rumen motility in sheep.[a]

Item	Frequency
	Cycles/min
Background, 45 dB	1.05 (.04)[b]
Continuous, 100 dB	1.14 (.05)
Intermittent, 100 dB	1.17 (.10)

[a]Continuous USASI white noise and pure intermitten sound
(7,000 Hz).

[b]Mean and standard error.   (Harbers et al., 1975)

individual's records, variation was great enough to mask any
effect sound stress might have inflicted. Observing the
animals tended to confirm that they reacted to sound only
during the first 10 to 15 minutes of each trial. The
recordings tended to become more irregular as the trial
went on, indicating perhaps a reaction to the digestion stall
environment.

In conclusion, the data confirm the finding of previous
studies (8,9,27) that animals probably acclimate to continuous and intermittent sounds of 100 dB or less. None
of the auditory stimuli apparently adversely affected digestibility or rumen motility in sheep exposed to short-
term experiments; intermittent sounds stimulated digestion.
Urinary creatinine values suggest that white noise and IMS
may reduce protein catabolism. The increased digestibility
coefficients produced when sheep were exposed to IMS sounds
(but not those produced when lambs were subjected to USASI
sounds) may partially explain the growth stimulation reported
in short-term studies (Experiment III).

EXPERIMENT V - ENDOCRINOLOGY AND REPRODUCTION

Biological assays of pituitary tissue from lambs exposed to different sound intensities (75 and 100 dB) in
Experiment II indicated altered levels of gonadotropin.
Thus, the initiation of a series of trials dealing with the
effect of sound on endocrine function.

As pointed out by Lockett (15), many stressors which increase ACTH reduce thyroid activity. To document sound as a stressor in large animals and to demonstrate the existance by hypothalamic relationships with auditory stimuli we measured thyroid activity in lambs exposed to sound. The trial involved 100 lambs subjected to 75 and 90 dB of white noise and compared to controls (62 dB background). Following 14 days of exposure serum samples were collected and analyzed for T-3% and PBI. T_4 index was calculated and is assumed to indicate circulating levels of thyroxine. A significant (P .01) decrease in T_4 index was noted in lambs exposed to 90 dB white noise (Figure 6) which supported findings of audio oriented changes in endocrine function, as well as an indication that sound is a stressor since decreased thyroid activity is an indicator of stress.

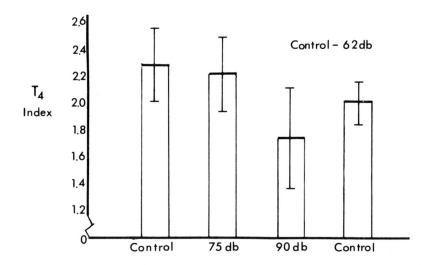

Figure 6 – T_4 index of lambs following 14-day exposure to 75 and 90 dB and return to control sound level (62 dB).

A trait associated with several stressors (electric shock, heat, restraint) has been the undesirable dark or light color beef and pork, respectively. Observations were made on color of two muscles (longissimus and rectus abdominus) of 42 lambs exposed to various sound types and intensities. Color was noted at 48 hours post slaughter by visual scores and reflectance spectrophotometry. Sound type did not affect visual score, but white noise caused more blue and less red reflectance. Intensity did not affect muscle color, but at 100 dB, music resulted in brighter visual color, but no clear response to type of sound was noted at 75 dB. Apparently, IMS and white noise is more stressful than music for lambs as indicated by post mortem muscle color.

Seventy-nine white faced ewes were used to evaluate the effects of sound type (USASI white noise vs. 4000 Hz pure tone), program technique (continuous vs. intermittent play), and time of exposure (day 12 to 17 vs. day 14 to 17 of estrous cycle) on ovarian structures and onset of estrus in synchronized ewes. Synchronization was accomplished using vaginal pessaries which were removed on day 14 of estrous cycle. All sound treatments were conducted at 100 dB intensity. On day 20, all ewes were laporatomized and ovarian structures characterized.

As shown in Table 6, audio environment significantly altered ovarian observations. It appears that 4000 Hz pure tone during days 14 to 17 (proestrus) are most effective in increasing number of copora lutentia per ewe. This observation bears directly on expected prolificacy of sound exposed animals. A logical explanation is that sound is a neurally interpreted stimulation which acts directly on midbrain centers via the auditory nerve much the same as light via the optic pathway. Hypothalamic integration in turn results in differences in gonodotropin releasing factors and, consequently, altered ovarian function. Follow-up studies comparing control ewes to sound-treated ewes (4000 Hz pure tone, day 14-17) have resulted in significant increases in number of lambs born (98.2 to 114%, respectively). In more recent experiments, sound treatments and exogenous injection of pregnant mare serum (PMS) at the rate of 750 IU have resulted in similar responses (Table 7). Kiracofe (personal communication) has found no changes in circulating levels of FSH or LH in ewes exposed to sound stress during anestrus. Tamari (24) in a series of experiments with laboratory rodents has reported inhibitory effects of auditory stimulation during pre-mating. In a study in-

TABLE 6

Effect of sound on ovarian measurements and estrus.

Sound type[1]	Program technique[2]	Day of exposure[3]	Follicle number	Corpora lutentia	Onset of estrus[4]
Control	---	---	5.67^a	1.11^a	36hr.
White	C	12-17	4.56^a	1.33^b	24hr.
White	C	14-17	5.77^a	$1.33^{b,c}$	12,24,36 hr.
White	I	12-17	4.00^a	$1.22^{a,b}$	24 hr.
White	I	14-17	6.00^a	1.56^d	24 hr.
4000 Hz	C	12-17	7.22^b	$1.25^{a,b,c}$	12hr.
4000 Hz	C	14-17	8.11^b	1.50^d	12 hr.
4000 Hz	I	12-17	7.78^b	$1.38^{c,d}$	12 hr.
4000 Hz	I	14-17	6.22^a	1.75^e	12 hr.

[1]All treatment exposures 100 dB intensity; controls 50 ± 5 dB.

[2]Continuous - C; Intermittent - I.

[3]Refer to day of extrous cycle with estrum being day 0.

[4]Interval of time after pessary removal when most ewes within a group exhibited estrus.

[a,b,c,d,e]Means in columns with different superscripts differ significantly ($P < .05$).

TABLE 7

Reproductive performance of ewe treated with sound and hormones.[1,2,3,4]

Group	Number	No. lambing to treatment	Lambing %
Control	20	2	150
Sound	20	6	130
Sound + HCG	20	9	166
PMS	20	8	150
PMS + HCG	20	12	166

[1] All groups were synchronized with vaginal pessaries.

[2] Sound treatment was 4000 Hz pure tone; intermittent play, day 14-17 of estrous cycle.

[3] HCG was injected at rate of 250 IU.

[4] PMS was injected at rate of 750 IU.

volving 40 ewe exposure to 10,000 Hz of 100 dB sound had no effect on conception rate or embryonic mortality. This failure of sheep to respond in a manner similar to laboratory rodents emphasizes the danger of predicting responses of large animals to auditory stimuli from data generated using laboratory rodents.

CONCLUSIONS

    Data presented here indicate that non laboratory animals are responsive to changes in the audio environment. It indicates that there is differentiation between sound type (varied frequency) and intensity. The data supports the premise of acclimation to sound. Results suggest a wide range of responses to sound with immediate physiological responses similar to those reported for laboratory rodents. However, the far reaching effects of the sound environment may, in fact, affect such basic functions as growth and reproduction.

REFERENCES

1. D.R. Ames. J. Animal Science. 1971. 33:247.
2. A.Anthony, E. Ackerman and J.A. Lloyd. J. Acoust. Soc. Amer. 1959. 31:1430.
3. A.O.A.C. Official Methods of Analysi - 11th Ed. 1970. Assoc. of Official Agri. Chemists. Washington, D.C.
4. L.A. Arehart. M.S. Thesis . Kansas State University. 1970.
5. L.A. Arehart and D.R. Ames. J. Animal Science. 1972. 35:481.
6. K.L. Blaxter. The Energy Metabolism of Ruminants. 1962. Hutchinson and Co. London.
7. J. Bond. Agr. Sci. Review. 1971. 9:1.
8. J. Bond, C.F. Winchester, L.E. Campbell and J.C. Webb. USDA Tech. Bull No. 1280. 1963. U.S. Dept. of Agric. Washington, D.C.
9. R.B. Casady and R.P. Lehman. National Sonic Boom Evaluation Office Interim Report NSBE-1-67. 1967.
10. L.H. Harbers, D.R. Ames, A.B. Davis and M.B. Ahmed. J. Animal Science. 1975. 41:654.
11. R.I.Henkin and K.M. Knigge. J. Physiol. 1963. 204:710.
12. K.A. Kemp. Research Paper 7. 1972. Kansas Agri. Expt. Stat. Manhattan, KS.
13. C.Y. Karmer. Biometrics. 1956. 12:307.
14. D.A. Laird. Med. J. and Rec. 1932. 135:461.
15. M.F. Lockett. In Physiological Effects of Noise, p. 23. 1970. Plenium Press. New York - London.
16. J.A. Owen, B. Iggo, F.J. Scondrett and C.P. Stewart. Biochem. J. 1954. 58:426.
17. O.L. Paladines, J.T. Reid, B.D.H. Van Niekerk and A. Bensadoun. J. Animal Science. 1964. 23:538.
18. J.B. Parker and N.D. Bayley. U.S. Agric. Research Service. 44-60. 1960. 60:22.
19. K.G. Pollack and F.C. Bartlett. Indust. Health Res. Bd. Rep. 65. 1932.
20. M.J. Riemann. M.S. Thesis. Kansas State University. 1973.
21. A.M. Sackler, A.S. Weltman and J.P. Jurtshuk. Acta. Endocrinology. 1959. 31:405.
22. R.G. Smith and D.A. Laird. J. Acoust. Society Amer. 1930. 94:98.
23. R.G. Steel and J.H. Torrie. Principles and Procedures of Statistics. 1960. McGraw-Hill Book Co. New York.
24. I. Tamari, In. In Physiological Effects of Noise, p. 117. 1970. Plenium Press. New York - London.
25. W.D. Ward and J.E. Trickel. Proc. of ASHA Conf., Amer. Speech and Hearing Assoc. Rep. 1960.

26.   B.L. Welch and A.S. Welch.   Physiological Effects of
      Noise.   1970.   Plenium Press.   New York - London.
27.   C.F. Winchester, L.E. Campbell, J.Bond and J.C. Webb.
      U.S. Wright Air Development Center Tech. Rep. 1959.
      59-200.

EASTERN WILD TURKEY BEHAVIORAL RESPONSES
INDUCED BY SONIC BOOM [1]

Thomas E. Lynch

Tennessee Wildlife Resources Agency
Ellington Agricultural Center
Nashville, Tennessee

Dan W. Speake

Alabama Cooperative Wildlife Research Unit
Auburn, Alabama

ABSTRACT

   *The nest sites of 8 to 20 wild turkey hens equipped with
164 MHz transmitters were located by telemetric triangulation,
and 4 of these were subjected to both real and simulated sonic
booms. Hens with young were subjected to simulated sonic
booms only. Sonic booms did not cause abnormal behavior that
would result in decreased productivity.*

INTRODUCTION

   The effect of sonic booms on native fauna has been discussed since the advent of supersonic flight. The possibility
exists that sonic booms might cause birds to abandon the nest
or cause the female to react in a manner that would disclose
the location of the nest to predators. It is also possible
that sonic booms might cause the female to desert her young,
or cause the young to scatter and become lost.

[1] *A contribution of the Alabama Cooperative Wildlife Re-
search Unit, which is jointly sponsored by Auburn University
Agricultural Experiment Station, Game and Fish Division of the
Alabama Department of Conservation and Natural Resources, the
U.S. Fish and Wildlife Service, and the Wildlife Management
Institute. The study was supported by a grant from the Fede-
ral Aviation Administration.*

47

The eastern wild turkey (<u>Meleagris</u> <u>gallopavo</u> <u>silvestris</u>) was chosen as the subject of this study because of its importance as a native North American game species and because of its shyness and the tendency of the hen to abandon her nest when disturbed by the activities of man (Leopold 1944, Wheeler 1948, Williams et al. 1970). The hearing ability of the wild turkey is well developed; Welty (1962) in describing the hearing ability of all birds stated, "Although the cochlea of the average bird is approximately only one-tenth the length of the mammalian cochlea, it has about ten times as many hair cells per unit of length. This shorter, broader construction of the hearing mechanism of the avian ear suggests to Pumphrey (1961) that birds are less sensitive to a wide range of sound frequencies than mammals, but more sensitive to differences in intensities. Further, a bird is able to hear and respond to rapid fluctuations in song about ten times as rapidly as man can. This fact is proved by the ability of young birds to imitate other birds' songs that have intricacies which are inaudible to human ears but visible in sound spectrographs".

The wild turkey not only has keen hearing but its vision is remarkable. The color perception and visual acuity of the wild turkey is equal to that of man but the rate of assimilation of detail, in the entire field of vision, is much higher than that of man; "Thus the vision of wild turkeys as a whole is no sharper but considerably faster than that of man" (Hewitt 1967). It was therefore important that investigations of the effect of sonic booms on wild turkey behavior be conducted in a manner that excluded the simultaneous occurrence of an unusual visual stimulus.

The wild turkey is a ground nesting bird, laying an average of 12.3 eggs per clutch (Hewitt 1967). One egg is laid per day and the completed clutch is then incubated for about 28 days. Only one clutch is hatched per year.

The nest site is usually in an area where the hen is concealed. This is important because the incubating hen remains on the nest almost continously, leaving it only briefly each day for food and water. Nesting normally begins in early April and 80% of the hatching is normally completed by mid-June in central Alabama (Speake et al. 1969).

After the eggs hatch, the hen usually leads her poults to an area adjacent to and including a pasture or a woods opening, where insects, grasses, and grass seeds heads can be found as food for the young. The first 10 days are critical because the poults cannot fly to safety when attacked by a predator. When a hen senses danger, she issues a "putt" sound, to which the poults react by remaining completely motionless until the hen signals that the danger has passed.

Often two or three hens with poults join together to form what is known as a brood group. This banding together offers survival advantages to the young by increasing the possibility that a predator will be quickly detected. The brood group remains together until late fall, when the young gobblers leave and form a separate flock.

The effect of sonic booms on the nesting behavior of wild turkey hens has not been investigated. Donohue et al. (1969), while observing an incubating wild turkey hen from a camouflaged blind, reported no change in the hen's behavior immediately after a sonic boom. Further observations of nesting hens, and also brood groups, at the time of sonic booms were justified to determine if abnormal behavior is induced which might result in losses of nests or poults. Normal nesting behavior and brood group behavior of the wild turkey have been described in several studies (William et al. 1968; Speake et al. 1969; Hillestad and Speake 1970; Williams et al. 1970).

THE STUDY AREA

The study was conducted at Saco, Alabama, in northeast Pike County and southwest Bullock Count of east central Alabama. Topography varies from flat to gently rolling stream bottoms, having an average elevation of 350 feet above mean sea level, to the steep upland ridges which average 500 feet above mean sea level.

The climate is temperate; summers are warm and winters mild. The area is sparsely populated and the principal industries are the production of cattle and timber.

PROCEDURES

Twenty wild turkey hens were captured during 6 - 24, March, 1973 - 15 by use of the oral anesthetic tribromoethanol (Williams et al. 1970), 3 by use of the oral anesthetic alpha-chloralose (Williams 1966), and 2 by use of a rocket-projected net (Dill 1969). In each of these capture methods, turkeys were first lured to a chosen bait site with whole-kernel corn. Cracked corn was then used as bait and the turkeys were allowed 1 to 2 weeks to become accustomed to feeding at the

bait site and to become somewhat dependent on this site for
food.   In late winter, choice naturally occurring wild turkey
foods are scarce.

A well-camouflaged blind was constructed 20 to 35 m from
the bait site and the number of turkeys using the bait site
was determined.   In capturing turkeys with tribromoethanol,
we used a dosage of 13 g of drug per cup of cracked corn.   We
provided one 118-ml portion of drugged corn for each turkey,
and the individual portions were 1 m apart.   The procedure
with alpha-chloralose was identical, except that we used 2 g
of drug per cup of cracked corn.   The use of a rocket-projected net involved the placement of bait in a 1 by 4-m strip,
1 m in front of the middle of a 20-m long by 10-m wide folded
net.   The net was propelled by three rockets, 61 cm long and
4 cm diameter and fuelled with a solid propellant.   The rockets were fired when the turkeys were feeding with heads down,
directly in front of the middle of the net.

Captured turkeys were transported individually in para-
ffin-coated cardboard boxes, 30.5 by 45.7 by 76.2 cm, to the
laboratory of the Alabama Cooperative Wildlife Research Unit
in Auburn, Alabama.   Here each bird was weighed, and aged by
examining the terminal primary wing feather and the greater
upper secondary covert patch, as described by Williams (1961).
Each turkey was then leg-banded with a numbered aluminum
band, and a 6.4 by 15.2-cm brightly colored vinyl wing marker
was attached to the patagium of both wings, as described by
Knowlton et al. (1964).

We then equipped each turkey with 164 MHz transmitters
weighing about 90 g, in the manner described by Williams
(1968).   The transmitters were attached "backpack fashion"
on the back of the turkey and secured with 3-mm inside-
diameter surgical turbing tied beneath the wings.   Transmitters were manufactured by Sidney L. Markusen Electronic
Specialties, Cloquet, Minnesota.[2]   The instrument on each
turkey had a different transmission frequency.   The turkeys
were returned to their holding boxes and, when fully recovered, were released at their capture sites, usually the
following day.

[2]*Reference to commercial products does not imply
Government endorsement.*

Individual hens were located regularly by telemetric triangulation (Cochran and Lord 1963). A 24-channel 164 MHz
portable receiver (also manufactured by Sidney L. Markusen
Electronic Specialties), with a hand-held yagi antenna, was
used. At least three compass bearings were taken on each hen
from different known points on a map and the locations of the
hens were plotted. After a hen was located at the same place
for several consecutive days and assumed to be nesting, the
investigator moved closer to locate the nest site. A permanent blind was then constructed after darkness near each
nest to observe the behavior of the hen during real or simulated sonic booms.

Supersonic overflights by military aircraft were flown
so that oscillograph recordings could be made to compare intensities of real (airplane-produced) sonic booms with those
of simulated sonic booms. Simulated sonic booms were produced by 5-cm mortar shells provided by the Federal Aviation
Administration which were launched from a 0.5-m length of
5.7-cm diameter polyvinylchloride pipe.

The measuring and recording equipment, supplied by NASA-
Langley Research Center, consisted of two microphone systems.
Each microphone was connected to a Dynagauge which was then
connected to two Burr-Brown amps. This assemblage of equipment provided a total of four inputs, which led to both an
oscillograph and a tape deck. The oscillograph provided on
light-sensitive graph paper an immediate evaluation of the
intensity of the sonic boom. The tape deck provided a record
of the signal from the Burr-Brown amps that could later provide an oscillograph print-out of the sonic boom, if needed.

The photocon microphone, normally a 20 Hz to 10 KHz instrument, consisted of a capacitor and a coil that together
composed a resonance tank circuit. The low-end frequency
response of 0.02 Hz to 10 KHz was achieved by restricting
the air flow of the microphone vent. The Dynagauge elec-
tronics provided the frequency required by the microphone
(about 710 KHz) and a tuner to sense changes caused by the
microphone. When the microphone diaphragm is exposed to a
pressure, the capacitance changes, which in turn causes the
resonance frequency of the tank circuit to change. This
change is sensed by the tuner and a voltage corresponding to
the microphone diaphragm motion is produced. The Burr-Brown
amplifier provided a gain of 0 to 60 dB in steps of 2 dB.
The flat frequency response is DC to 20 KHz or more. The
galvanometer amplifiers provide up to 100 mA of current to
drive the galvanometers in the direct-write paper recorder.

The tape recorder is frequency modulated, operating at 30 inches per second in the intermediate IRIG band with center frequency of 54 KHz that provides a frequency response of DC to 10 KHz.

The two microphones were mounted side-by-side on a 1.2-m square sheet of 1.3-cm thick plywood, and covered by a wind sock. They were then connected by 304.8 m of coaxial cable to the Dynagauges. Electrical power for this equipment was produced by a portable gasoline-powered generator. The equipment was mounted in a 4-m mobile aluminum camper-trailer so that it could be easily transported in the field.

To accurately measure the intensity of the sonic booms that nesting wild turkey hens were being exposed to, we set up the microphones and recording equipment in an open, flat pasture, bordered by forested, gradually rising hillsides. A total of five real sonic booms--one on 29 May 1973 (at 1712 hours) and four on 30 May (one each at 0823, 0839, 1607, and 1629 hours)--were produced by military aircraft in cooperation with the Federal Aviation Administration. The behavior of each of four individual nesting hens at the time of each real sonic boom was observed from camouflaged blinds and described in detail by personnel of the Alabama Cooperative Wildlife Research Unit.

In subjecting nesting hens to simulated sonic booms, the investigator approached the blind slowly and entered it cautiously, to avoid alerting the hen. After waiting quietly at least 30 minutes, the investigator, using binoculars, visually located the incubating hen on the nest. When satisfied that the hen was not aware of his presence, he signalled by radio an assistant, who then launched a mortar shell from a point 300 to 500 m away, well out of sight of the hen. The intensity of the sound produced by the shell was comparable with that provided by real sonic booms.

While subjecting brood groups to simulated sonic booms, the investigator waited in a camouflaged blind in an area used by the groups. When a group was sighted, he signalled by radio an assistant, who then launched a mortar shell, again from a point 300 to 500 m away, well out of sight of the brood group.

MEASUREMENT AND ANALYSIS OF SONIC BOOM PRESSURE

A typical record for one of the airplane flyovers is shown in Figure 1 (following page).

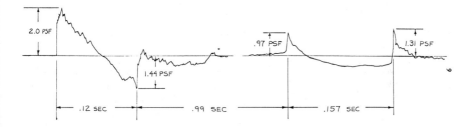

Figure 1 – Typical recording of the pressure (measured in
          pounds per square foot) and the time span between
          shock waves (measured in seconds) of an airplane-
          produced sonic boom at Saco, Alabama, on 30 May
          1973.

This is a time history of the pressure recorded by one of
the microphones. The magnitude of each shock wave is indicated, as is the time span between each. This particular
record consists of two distinct pressure signatures following
one another. The signatures are characteristic of those produced during a level flight acceleration of the airplane
(Kane 1973). The leading signature is generally N-shaped
and the following signature is usually U-shaped. At some
location prior to the position of this recording (further
up the flight track toward the origin of the flyover) the
two signatures were merged, producing a single signature with
stronger shock waves.

     Observations of sonic booms simulated with a 5-cm mortar
shell are tabulated in Table 1 (following page) for the recordings made on 31 August 1973. In a typical set of signatures (Figure 2) the time delay between shock waves of shell
bursts is much shorter than that for the airplane flyover.

Table 1. Summary of recordings of the pressure (measured in pounds per square foot) and the time span between shock waves (measured in seconds) of twenty 5-cm mortar shells launched at Saco, Alabama, on 31 August 1973.

| Run | P_1 | | | | Avg. P_1 | Avg. P_2 | T |
	Mic.1	Mic.2	Mic.3	Mic.4			
1	0.746	0.818	0.868	-----	0.811	0.673	0.0050
2	0.617	0.662	0.681	0.691	0.663	0.537	0.0045
3	1.119	1.051	1.140	0.791	1.025	0.794	0.0045
4	0.783	0.799	0.834	0.759	0.794	0.703	0.0050
5	0.746	0.701	0.706	0.719	0.718	0.516	0.0053
6	0.821	0.818	0.851	0.850	0.835	0.594	0.0055
7	0.746	0.721	0.740	0.764	0.743	0.444	0.0060
8	0.448	0.428	0.434	0.452	0.441	0.346	0.0048
9	0.858	0.749	0.834	0.818	0.815	0.718	0.0061
10	0.821	0.760	0.817	0.795	0.798	0.674	0.0048
11	0.783	0.779	0.766	0.786	0.779	0.606	0.0048
12	1.007	0.974	0.885	0.949	0.954	0.708	0.0040
13	0.783	0.701	0.689	0.700	0.718	0.622	0.0053
14	0.932	0.935	0.919	0.646	0.858	0.752	0.0048
15	1.193	1.032	1.089	0.741	1.014	0.805	0.0050
16	0.821	0.701	0.749	0.696	0.742	0.567	0.0053
17	0.970	0.896	0.936	0.719	0.880	0.555	0.0045
18	0.895	0.818	0.868	0.637	0.804	0.721	0.0043
19	0.597	0.545	0.545	0.547	0.559	0.685	0.0058
20	0.597	0.565	0.612	0.610	0.596	0.442	0.0050

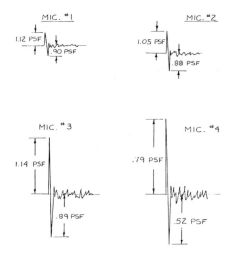

Figure 2 - Typical recording of the pressure (measured in
           pounds per square foot) and the time span between
           shock waves (measured in seconds) of a 5-cm mortar
           shell launched at Saco, Alabama, on 31 August 1973.

As a result, the human ear perceives the stimuli as a single
rather than a double boom. The magnitude of the shock waves
shown in Figure 2 is different for each microphone because
the run levels set for each microphone and channel in the
recording equipment differed.

    The magnitude of the overpressures recorded from the
airplane produced (real) sonic booms are typical of those
generated directly beneath the flight path by current supersonic transports such as the British-French Concorde and the
Russian TU-144 (Anonymous 1970, 1973). To the side of the
flight track (a tracing of the flight path upon the ground)
the sonic boom intensity diminishes. The area on the ground
underneath the airplane flight path that is exposed to the
sonic boom noise is called the "sonic boom carpet" and extends
a finite distance to the sides of the flight track. Near the
edges of the carpet, the magnitude of the overpressure reaches

a value near 1.0 pounds per square foot which is comparable
to the intensity of the simulated sonic booms. Hence, the
birds were exposed to nearly the full range of sonic booms
that would be generated during supersonic transport flight.

RESULTS

The nest of Hen A, containing eight eggs, was found on
5 May 1973. The hen was subjected to eight simulated sonic
booms from 12 May to 21 May 1973, while she was on her nest.
No unusual effects were observed (Lynch and Speake 1975,
Table 1).

The reaction of this hen at the time of a simulated sonic
boom was as follows: the hen appeared relaxed on the nest
with head held low and somewhat close to her body. At the
sound of the boom, the hen quickly lifted her head 5-15 cm
and appeared alert for 10-20 seconds but did not move her
head about. She then lowered her head and once again appeared relaxed. The hen did not rise off the nest, flap
her wings, call out, or do anything to disclose the location
of her nest.

On 22 May 1973, two eggs had hatched and one infertile
egg was found in the nest. No evidence of the other five eggs
could be found; they were presumably removed by predators.

No observations of the behavior of this hen at the time
of a real sonic boom were made because hatching had occurred
before sonic booms were begun.

The nest of Hen B, containing eight eggs, was found on
12 May 1973. This hen was subjected to six simulated sonic
booms and five real sonic booms from 20 May to 11 June 1973,
while she was at rest on her nest. No unusual effects were
observed.

This hen reacted in the same manner as Hen A to both real
and simulated sonic booms. On 13 June 1973, the shells of
eight eggs that had hatched were found in the nest.

The nest of Hen C, containing twelve eggs, was found on
14 May 1973. This hen was subjected to four simulated sonic
booms and five real sonic booms from 20 May to 30 May 1973,
while she was at rest on her nest; no unusual effects were
observed.

This hen also reacted in the same manner as Hens A and B to both real and simulated sonic booms. On 5 June 1973, all twelve eggs had hatched.

The nest vicinity of Hen D was located on 27 May 1973. Because three hens had abandoned their nest before being subjected to sonic booms as a result of activities of the investigator, and because no reactions to sonic booms had been observed in Hens A, B, and C, Hen D was not observed directly. The investigator positioned himself so that any sudden movements by the bird on the nest could be visually detected. Before each real or simulated sonic boom, the location of the hen on the nest was checked by use of the telemetry equipment. This hen was subjected to six simulated and five real sonic booms from 29 May to 13 June 1973; no unusual effects were observed.

On 14 June 1973, it was discovered that the nest of Hen D had been destroyed by a predator. The eggs were broken and the contents consumed, and many of the hen's feathers were scattered about the nest. The investigator immediately located the hen by using the telemetry equipment and visually determined that she had survived the attack; she was observed on many occasions throughout the summer.

REACTIONS OF BROODS TO SIMULATED SONIC BOOMS

Twenty-one brood groups were observed as they were subjected to simulated sonic booms. One of these was not considered valid because the brood group was aware of the presence of the investigator.

The reaction to simulated sonic booms in ten (50%) of the observations was as follows: the brood group was feeding undisturbed. At the sound of the launch blast the hens and most of the poults "stood at attention". Seven seconds later, at the sound of the simulated sonic boom, the hens and poults turned and ran toward the woods for a distance of 4-8 m and then abruptly stopped. The poults began to feed again immediately while the hens "stood at attention" and looked about. The hens remained alert for 15 to 20 seconds and then began to feed with the poults. Feeding and behavior then continued as it had before the boom.

The reaction to simulated sonic booms in six (30%) of the observations was as follows: at the sound of the launch blast there was no indication of alarm by the hens or poults. At the sound of the simulated sonic boom, the hens and poults "stood at attention" and looked about. After approximately

3-5 seconds, the poults began to feed again. The hens "stood at attention" and remained alert for 10-20 seconds after the boom and then began to feed with the poults. Feeding and behavior then continued as it had before the boom.

In two of the observations the following reaction occurred: at the sound of the launch blast all of the hens and about half of the poults "stood at attention". The poults then began to feed again and continued to feed even at the sound of the boom. The hens remained alert for 10-20 seconds after the boom and then began to feed with the poults. Feeding and behavior then continued as it had before the boom.

In one instance at the sound of the launch blast the hens and poults "stood at attention". At the sound of the boom, one hen jumped about 0.5 m into the air and then, almost immediately, the rest of the hens and poults also jumped. The brood group was feeding in a pasture and was 7-12 m from the edge of the woods. After the birds had jumped, the brood group hurried into the edge of the woods, where the poults began to feed again immediately. The hens remained alert for 10-15 seconds and then began to feed with the poults. Feeding and behavior then continued as it had before the boom.

In one instance the hens and poults also "stood at attention" at the sound of the launch blast. At the sound of the boom, the hens began to move from the pasture toward the woods at a fast walk while the poults fed as they moved along, also at a fast walk. Upon reaching the woods the hens immediately began to feed with the poults and behavior then continued as it had before the boom.

In no instance did the hens desert any poults. None of the poults scattered and became lost from the rest of the brood group. In every observation the brood group resumed its normal activities within a maximum of 30 seconds after a simulated sonic boom.

CONCLUSIONS

The results of this study indicate that sonic booms do not initiate abnormal behavior in wild turkey that would result in decreased productivity.

ACKNOWLEDGMENTS

We thank Joseph K. Powers and Tom Higgins of the Federal
Aviation Administration; Dave Hilton and Herb Henderson of
NASA-Langley Research Center for their advice and guidance;
Thomas Baxter of Wyle Laboratories, Hampton, Virginia, for his
assistance in the operation and set-up of the recording
equipment; Edward J. Kane of Boeing Commercial Airplane
Company, Seattle, Washington, for writing the section en-
titled "Measurement and Analysis of Sonic Boom Pressure" and
for his critical review and corrections of the final report;
Charles Kelley, Director of the Game and Fish Division
of the Alabama Department of Conservation and Natural Resources for furnishing the Division's airplane and pilot for
use in locating hens with which we had lost radio contact;
and J. B. Money and Joe Burt of Saco, Alabama, for permission
to conduct this study on their lands and for their cooperation
and encouragement.

LITERATURE CITED

1. Anonymous. 1970. Sonic boom panel - second meeting. ICAO Document 8894, SBP/II.
2. Anonymous. 1973. Sonic boom committee - second meeting ICAO Document 9064, SBC/x.
3. Cochran, W.W., and R.D. Lord, Jr. 1963. A radio-tracking system for wild animals. J. Wildl. Manage. 27(1):9-24.
4. Dill, H.H. 1969. A field guide to cannon net trapping. U.S. Bureau of Sport Fisheries and Wildlife, Washington, D.C. 18 pp. (mimeo).
5. Donohoe, R.W., C.E. McKibben, and C.B. Lowry. 1968. Turkey nesting behavior. Wilson Bull. 80(1):103.
6. Hewitt, D.H., (ed.) 1967. The wild turkey and its management. The Wildlife Society, Washington, D.C. 589 p.
7. Hillestad, H.O., and D.W. Speake. 1970. Activities of wild turkey hens and poults as influenced by habitat. Proc. Annu. Conf. Southeast. Assoc. Game Fish Comm. 24:244-251.
8. Kane, E.J. 1973. Review of current sonic boom studies. Aircraft 10(7):395-399.
9. Knowlton, F.F., E.D. Michael, and W.C. Glazener. 1964. A marking technique for field recognition of individual turkeys and deer. Wildl. Manage. 28(1):167-170.
10. Leopold, A.S. 1944. The nature of heritable wildness in turkeys. Condor 46(1):33-97.
11. Lynch, T.E., and D.W. Speake. 1975. The effect of sonic boom on the nesting and brood rearing behavior of the eastern wild turkey. Report No. FAA-RD-75-2. National Technical Information Service, Springfield, Virginia.
12. Pumphrey, R.J. 1961. Sensory organs: hearing. In A.J. Marshall, Biology and comparative physiology of birds. Academic Press, N.Y.
13. Speake, D.W., L.H. Barwick, H.O. Hillestad, and W. Stickney. 1969. Some characteristics of an expanding turkey population. Proc. Annu. Conf. Southeast. Assoc. Game Fish Comm. 23:46-58.
14. Welty, J.E. 1968. The life of birds. W.B. Saunders Co., Philadelphia and London.
15. Wheeler, R.J. 1948. The wild turkey in Alabama. Ala. Dep. Conserv. Bull. 12. 92 pp.
16. Williams, L.E., Jr. 1961. Notes on wing molt in the yearling wild turkey. J. Wildl. Manage. 25(4):439-440.
17. Williams, L.E., Jr. 1966. Capturing turkeys with alpha-chloralose. J. Wildl. Manage. 30(1):50-56.

18. Williams, L.E., Jr., D.H. Austin, N.F. Eichholz, T.E.
     Peoples, and R.W. Phillips. 1968. A study of nesting
     turkeys in southern Florida. Proc. Conf. Southeast.
     Assoc. Game Fish Comm. 22:16-30.
19. Williams, L.E., Jr., D.H. Austin, T.E. Peoples, and
     R.W. Phillips. 1970. Capturing turkeys with oral
     drugs. Proc. Natl. Wild Turkey Symp. 2:219-227.

EFFECT OF SONIC BOOM FROM AIRCRAFT ON WILDLIFE
AND ANIMAL HUSBANDRY

Ph. Cottereau

National Veterinary School of Lyon
Lyon, Cedex I, France

INTRODUCTION

The sonic boom is a new characteristic of the outside surroundings, and is thought to be able to have a repercussion on organisms.

The introduction of commercial and military supersonic aircraft has raised the question of whether booms are to be considered as severe environmental pollution with adverse effects on humans animals and structures. Much of present knowledge is based on occasional booms, many of which have resulted in complaints and claims. Although probably not always legitimate, these complaints indicate that concern has developed about the effects of the new environmental factor and this concern should stimulate intensified research. However, only a few investigations under real or simulated conditions have been undertaken so far in order to try and elucidate the possible effects.

Behavioural responses to sonic booms in domesticated animals such as horses, cattle, sheep and poultry are dealt with in a small number of papers. These species, plus mink, are often mentioned in the claim files (Grubb et al. 1967). Whether the studies have been performed with real or simulated booms, the authors come to the same general conclusions: sonic booms and subsonic flight noise have very little effect on the animal's behaviour (Fausst 1966, Boutelier 1967, Casady and Lehmann 1967, Nixon et al. 1968, Cottereau 1972, Espmark 1972, Ewbank 1977).

Avian species seem to be more affected than mammals. It is also suggested that the animals develop adaptation to the disturbances. The pressure waves produced by supersonic aircraft cause variation of pressure on and in the ground which are audible as the sonic boom. The noise can influence the behaviour of farm and wild animals while the variation of pressure in itself could eventually cause physiopathological disorders.

In all countries overflown by military supersonic aircraft complaints have been encountered from farmers and ranchers. Most reactions reported were changes in behaviour (fright. flight, etc.) which could result in secondary injuries.

In many countries, research has been undertaken to study the exposure effects on different species of animals through the use of aircraft or generators of sonic booms. A review of animal studies was presented by Bell. He also compiled the complaints received by the U.S. Air Force in terms of animal species each year from 1961 to 1970.

FIELD STUDIES BASED ON REAL AND ARTIFICIAL SONIC BOOMS

Hatchability of Eggs

The evaluation of effects of sonic booms on the hatchability of chicken eggs was studied at White Sands Proving Grounds. The particular set of eggs exposed for the full period received over 600 booms. It has a hatch of 84-3 % (control 84-2 %). All exposed sets had a mean hatch of 83-2 % and the unexposed control 81-3 %. No developmental deviations were found in sample eggs examined during the test, and no gross pathology was found in birds necropsied at 12 weeks of age. These birds were all normal in size and weight.

Robertson described an extensive hatching failure of Dry Tortugas Sooty Terns. Two reasons as to the effects of the booms were suggested: death of the embryos after abandonment of the colony by the terns in panic flight after exposure, or physical damage to eggs not covered by a sitting bird at the time of the boom.

Reactions in Farm Animals

Investigations on effects of sonic boom (average 72/Nm2, maximum 265 N/m2) on farm animal behaviour and per-

formance were made in 1966 at Edwards Air Force Base in
California.   The animal population consisted of horses, beef
cattle, turkeys, broilers, sheep dairy cattle and pheasants.

Except for the avian species the behavioural reactions
to the sonic booms were considered minimal.   Occasional
jumping, galloping, bellowing, and random movement were
among the effects noted.   The responses of the large farm
animals in these tests were judged to be in the range of
normal activity in comparison with animals observed under
controlled conditions.   The poultry observed showed more
response than the large animals, especially in the early
stages of the tests.   Occasional flying, running, crowding
and cowering were noted.

Additional observations of reactions of livestock to
sonic booms reported by Nixon et al. largely confirm the
above observations.   Espmark et al. (1974) give an account
of the behavioural responses of dairy cattle and sheep
exposed to a number of sonic booms of between 100 and 400
N/m2 overpressure and with rise times of 0,1 - 12 milliseconds (average 2,5) produced by Swedish Draken fighter
planes, and state that no adverse effects were observed
and that behavioural reactions were minimal.

Cattle

Cattle are generally described as briefly (i.e.
usually for less than 1 minute) stopping their current
behavioural activity and (or getting up and) or moving
several steps away from and or orientating toward the
direction of origin of the sound.   There seem to be no
reports of continued arousal or general panic.   Milk yield
does not seem to be affected (Casady and Lehmann 1967 -
one herd of 350 cows does not seem to be altered (Bond et
al. 1974).

There were neither deleterious changes in behaviour
nor alterations in semen quality or quantity when bulls
at an artificial insemination centre were subjected to
simulated booms (Cottereau, Bastien et al. 1973).

Pregnant Charollais beef cows were not alarmed and
their calves were born at term and were normal after exposure, during the first month of pregnancy, to 20 concorde-like simulated booms (Cottereau, Bastien et al. 1973).

There do not appear to have been any studies of the
possible effects of sonic booms on the actual milking of
cows.

Figure 1 - Artificial sonic boom and semen collection
(Concorde-like simulated boom).

Sheep

Sheep have been described as temporarily stopping
feeding, grazing or ruminating and (or getting up and)
or running together in response to sonic booms. There
appear to be no reports of panic, injury or impaired reproduction.

Pigs

There are few accounts of the effects of sonic booms
on this species. Pigs being kept both in building and
outside are reported to briefly stop what they are doing
and to remain quiet for a number of seconds. There have
been no indications of economic loss.

Horses

It has been suggested that horses, when compared with
the other grazing-species, may show a more violent response to impulse noises. A few have been reported as
showing muscular tremors, galloping and jumping, while

others seem to be much more affected. Ponies have been
described (Bond and al. 1974) as reacting to a greater extent than beef cattle when exposed to the same double
simulated booms (overpressures of approximately 200 N/m2),
but even their responses were mild. Cottereau, Chavot et
al. (1973) suggest that horses being ridden are easily
controlled when startled by simulated Concorde-type sonic
booms, but that they do not necessarily adapt readily to
them. There is a brief suggestion (Report 1973) that
thoroughbreds do not habituate, or habituate to only a
slight degree, when compared with half-breds. There is
the possibility that horses confined in a building may
show an exaggerated response as a result of either feeling
trapped and thus over-reacting and/or being alarmed by
the vibration of the building around them.

Poultry

Most comparative accounts of the behaviour shown by
domesticated animals in response to sonic booms, real or
simulated, emphasize the pronounced reactions of poultry
when contrasted with the farm mammals.

The birds, whether they were inside buildings or outside, have been variously described as agitated and/or
going quiet. A part from a small number of disputed claims
from Hatcheries there seem to be few suggestions of actual
loss.

Cottereau conducted field experiments with a mobile
sonic boom simulator. Comparisons were made between the
respective behaviours of broilers in two poultry farms, one
of them exposed to sonic booms (100 N/m2 300 ms duration).
Tests were made on the growth of broilers during breeding
and on laying. Results show that broilers which have been
exposed since their birth to sonic booms at a rate of three
every morning every day show a startle reaction and stop
all activity for 20 to 30 s.

Studies have been conducted where the effects of sonic
booms on chick embryos during hatching are studied in terms
of embryo death or congenital malformation. Experiments
were carried out with increasing boom intensity (100 N/m2,
300 ms duration, 500, 300; 1 500, 300; etc.). Chick embryos in hatching which were exposed to three sonic booms
every morning and three every evening every day (100 N/m2,
300 ms, 500, 300; 1 500, 300; 3,000, 300) were not affected.
Chicks from these eggs were normal.

Fertile eggs in incubators have been exposed to over-pressure of up to 5,000 N/m2 (i.e., 50 times the over-pressure from civil supersonic transport aircraft) with rise times of less than 100 milliseconds without deleterious changes being seen in the eggs or embryos (Report 1973), and Cottereau, Place et al. (1973) provoked mild startle reactions in laying poultry by using simulated Concorde-like sonic booms but found that the birds soon adapted and that there were no deaths or drops in egg production due to the treatment the birds received.

Reactions of Pets

Boutelier reported on some observations and results of his experiences in France with French Army dogs subjected to sonic booms with high overpressure during focalization. A minor effect on the cardiac frequency and on the dog's behaviour was found. Certain dogs became frightened, fled and sometimes became aggressive. The effect of simultaneous training of the dog varied: the more the work required attention from the dog the less it was affected by the boom.

Reactions of Mink

Mink reactions to sonic booms have received more attention than reactions of other animal species possibly because these animals are known to be rather sensitive to unusual sounds. An exploratory field study was conducted in 1967 on a mink ranch in the centre of a supersonic flight test corridor in Minnesota. The responses to the booms were reported as similar to responses to truck traffic, snow plows, barking dogs and mine blasting noise in the area.

Travis et al. reported on another study of mink performed in the spring of 1967. The general conclusion were that differences in boom intensities (average 72 N/m2, maximum 96 N/m2) had no effect, that repeated booming produced no signs of increased excitability and that reproduction in boomed and non-boomed mink was normal.

An interdisciplinary study was conducted by Bond on the effects of real and simulated sonic booms (average 242 N/m2, maximum 314 N/m2) on the behaviour, reproduction and growth of farm-raised mink (Mustelavison). The study was conducted on Mitkof Island, near Petersburg, Alaska. The booming of the mink was conducted in the middle of the whelping season. Observers hidden behind screens studied the reaction of the animals. In addition, films were taken before and after the exposure. Only minimal or no disturbance was noted and frantic activities or alarm screams were not observed.

Reactions of Wild Animals

Hubbard reported that wild deer studied at Eglin Air Force Base showed no apparent response to high level sonic booms.

Rawlins reported that the animals in the London Zoo were observed in 1968 during a short program of sonic booms over London. Except for a small group of young chimpanzees, which showed a tendency toward fright, the reactions of the zoo animals were negligible.

Espmark reported results from a sonic boom field test conducted in northern Sweden 1970 to study the response of reindeer confined in a one-acre corral directly under the flight path. Sonic booms of around 60 N/m2 caused a slight startle effect among the herd of reindeer, but only ex-ceptionally were on-going activities interrupted. As boom levels increased up to 200 N/m2 the reactions became more noticeable but none of the lying or resting animals arose.

Davies has described ravens' responses to a sonic boom in the English countryside. On one occasion three or four ravens were noted flying idly in the up-currents over a rocky spur when a sonic boom was generated. Im-mediately ravens began to call and to fly in from several directions to converge over the spur. In about 5 min. 50 or more ravens were present, flapping, soaring and chasing each other. Some settled briefly and noisily on the rocks. Within 10 min. they began to disperse and the calling died down considerably.

Ruddlesden (1971) could not demonstrate a fall in egg production when a pheasantry was exposed to frequent impulse noises from an explosive device. The general locality, however, was subjected to frequent impulse noises from an explosives testing range and thus the nil response of the pheasants may have been the result of a prolonged period of adaptation. Teer and Truett (1973) were unable to alter the hatching success, growth rate or mortality of Bobwhite quail by generating simulated booms of 100-250 N/m2 close to the incubators and/or young birds.

Many studies have been performed to determine the sensitivity of fish to sounds, principally using conditioning procedures. Though these have indicated that fish are extremely sensitive to low frequency sounds little information is available on the normal response of these animals to either naturally occuring or man-made sounds. Experiments have been performed on the sensitivity of cod to sound stimuli where the response was measured by an electrocardiographic technique.

Only one sonic boom has been examined in detail. In this case, a boom of an approximate overpressure of 100 N/m2 gave an over pressure of 1 N/m2 at a depth of 15 m in the sea. The rise-time of the boom was less pronounced in the water than in air. Despite the great reduction in amplitude-arising from the acoustical mismatch between air and water-a single fish did show a brief slowing of heart-rate (bradycardia), immediately after the arrival of the boom. However, fish frequently respond similarly to other sound stimuli, and in particular to sounds generated by ships. It should also be stressed that the sea is a naturally noisy medium, and that sound pressure amplitudes of this level are not uncommon, especially in busy harbours, or close inshore. It would therefore seem that sonic booms are unlikely to have any serious effects upon fish, though it is clear that they can be detected by the animals and may produce a brief startle response. Further research is warranted in this area.

SONIC BOOM

Physical Aspects

A plane travelling at a speed greater than the speed of sound generates pressure waves as its wings, tail and body pass through the air, but these waves will not be able to escape forward from the aircraft as it (the plane) is travelling faster than they are. This sets up conditions by which high energy sound waves are produced and these then, moving out from the plane at the speed of sound, in effect form a cone-shaped sonic boom front behind the aircraft (Figure 2 and 3).

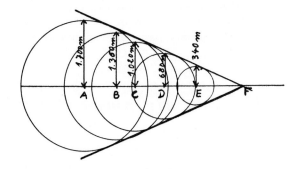

Figure 2 - A cone-shaped sonic boom.

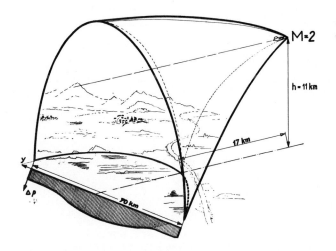

Figure 3 - Right supersonic flight: sonic boom carpet.

The loudness of these high energy sound waves is only reduced by a factor related to the three-fourths power of the distance travelled and not, as in the low engery sub-sonic produced waves, to the inverse square of the distance.

In the Figure 3, which represents the situation at one instant of time, the aircraft is trailing behind it a cone-shaped shock wave.

This is a rough drawing of an oscillograph trace which has plotted out pressure against time for one sonic boom. This type of diagram is sometimes referred to as the sonic

boom signature. There is a sudden build-up (N-M) of overpressure above atmosphere pressure which, in the case of
Concorde travelling at 1,3 Mach at a height of ca 15,000
m is about 100-125 newtons par metre squared [1] (Normal
atmospheric pressure is of the order of 100,000 N/m2).

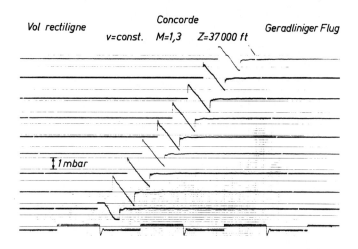

Figure 5 - Oscillograph tracing of "Concorde" sonic boom.

The greater the speed of the aircraft and the lower its
altitude the greater is the overpressure, and readings of
up to 500 N/m2 have been recorded from fast, low flying
military aircraft.

These degrees of pressure are especially seen when
planes accelerate into supersonic flight and when they
bank and in effect focus the boom on one place. Certain
temperatures and wind patterns in the air may at times
also keep to focus and exaggerate a sonic boom. These
exaggerated or superbooms can be very local and only
affect a relatively small area.

[1]The Newton per square metre (or Pascal) is the MKs
system unit of pressure. One newton is the force that will
accelerate 1 kg at 1 metre per second: 48 N/m2 = Ilb/ft^2
i.e., 100-125 N/m2 2-2,5 lb/ft^2. Atmospheric pressure is
of the order of 2,000 lb/ft^2 i.e., the pressure fluctuations
due to a normal sonic boom from Concorde is about 1/1,000
of the atmospheric pressure.

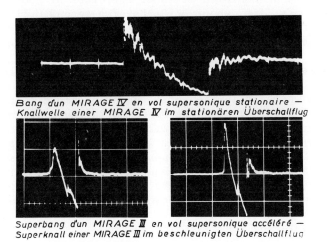

Bang d'un MIRAGE IV en vol supersonique stationaire —
Knallwelle einer MIRAGE IV im stationären Überschallflug

Superbang d'un MIRAGE III en vol supersonique accéléré —
Superknall einer MIRAGE III im beschleunigten Überschallflug

Figure 6 – Boom and superboom of "Mirage" aircraft.

The rise time (t) of normal booms is usually short
and of the order of 2-3 milliseconds. The sharper the rise
time, the sharper the noise, i.e. it becomes more a crack
or bang than a boom.

This first pressure rise is followed by a drop to below
atmospherique pressure and then a sudden rise again (R-S)
of some 100-125 N/m2. The time T (Figure 4) between the onset of the two positive pressure changes is known as the
signal interval and its duration is a function of the aircraft length.

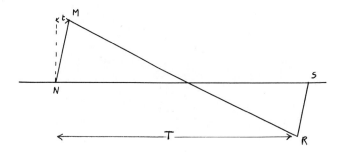

Figure 4 – Characteristics of a sonic boom – Pressure
          (vertical axis) plotted against time (horizontal
          axis).

In the case of Concorde it is of the order of 250-300
milliseconds, i.e. about 3/10 ths of a second. If the
signal interval is greater than 1/10 th of a second it will
usually be perceived as two separate noises.

Most supersonic fighter planes have signal intervals
of about 100 milliseconds. The term sonic boom really
covers a whole collection of N shaped pressure (sound)
waves (signatures) with the shape as in the diagram and
with short rise times (1-10 milliseconds) in the range
50-500 N/m2 overpressures and with signal intervals usually
between 50 and 300 milliseconds. The actual values of
these components will vary according to type of aircraft,
its speed and load, its height above ground, atmospheric
conditions, etc. From the point of view of their effect
on animals the important factors would seem to be the
overpressure and the rise time.

Boom, Generators and Simulators

Simulators have been developed which can be used to
expose animals in their native environment.

Figure 7 - French mobile sonic boom generator.

Figure 8a – Sonic boom generator of the French and German
Institute of Saint-Louis (France).

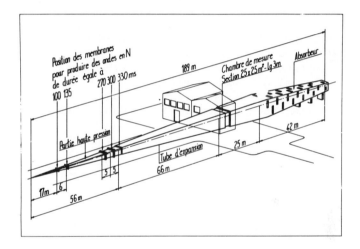

Figure 8b – Sonic boom generator diagram of the Saint-Louis,
France generator.

CONCLUSION

It appears that all the domesticated farm and wild
animals show behavioural startle when they first experience
a sonic boom. Their reaction is usually slight and they
seem to adapt readily to further booms.

Poultry are reported to give the most pronounced responses. It has not been possible as yet, to clearly demonstrate that deaths, injury or loss of production have
arisen as a consequence of sonic boom exposure.

In the early days of supersonic flight a number of claims for compensation for damage to animal enterprises were made, and it now seems settled more for the sake of convenience than as a result of clear-cut evidence.

The greatest research need is for critical observation of the response of aggregations of various social mammals and birds to sonic booms of measured overpressure and duration. Cooperative research in this area is highly recommended with a large participation of biologists. Concerning wild animals the following statements may be made with a degree of confidence effects upon wild mammals or birds; sonic booms of extreme overpressures (20,000 N/m2) may have the potential to crack bird eggs, but the question awaits an adequate experimental test: chronic, direct effects on wild animals have not been investigated, but no significant effects of this kind are presently forseen; the disturbance effects of sonic booms of overpressure around 100 N/m2 on wild mammals are probably insignificant, but the responses of a much greater diversity of species should be studied. Such booms are markedly more disturbing to birds than to mammals, and, in the case of colonially nesting birds, adverse effects may result from sonic boom exposures.

REFERENCES

1.   Bell, W.B. 1972. Journal of the Acoustical Society
      of America (in press). Animal response to sonic booms.
      Paper presented at the Symposium on Sonic Boom, Annual
      Meeting of the Acoustical Society of America, Houston,
      Texas, November 3-6, 1970.
2.   Bond, J. 1970. In Physiological Effects of Noise.
      New York: Plenum Press, pp. 295-306. Effects of noise
      on the physiology and behaviour of farm-raised animals.
3.   Bond, J. 1971. Personal communication at Sonic Boom
      Workshop (to be published).
4.   Bond, J., Rumsey, T.S., Menear, J.R. et al.(1974):
      In Proceedings of International Liverstock. Environment
      Symposium Lincoln, Nebraska. Special publication SP-
      0174 of the A.S.A.E., p. 170.
5.   Boutelier, C. (1967) Revue Corps Vet. Biol. Armees (F),
      20, 112-117, (Lab. Med. Aerospatiale, Bretigny sur Orge
      Now listed as, Fr. Service Biologique et Veterinaire
      des Armees. Revue. Original not available. Material
      cited from Veterinary Bulletin 38 (1968) Abstract
      1986, The sonic bang, its effect on man and animals.
6.   Casidy, R.B. and Lehmann, R.P. (1967): Studies at
      Edwards Air Force Base. June 6-20, 1966, Interim
      Report, Section H, National Sonic Boom Evaluation
      Office, Arlington, Virginia Contract AF49 (638)-1758:
      Stanford Research Institute. Response of farm animals
      to sonic booms.
7.   Cottereau, Ph. and Keller, M. (1971): Ecole
      Nationale Veterinaire Lyon France. Les effets du bang
      supersonique sur certains animaux.
8.   Cottereau, Ph., Bastien, G., and Place, Ch. (1973):
      Rapport d'Experimentation sur les effets du bang soni-
      que, IV Lyon,Ecole Nationale Veterinaire.
9.   Cottereau, Ph., Chavot, L., Martin, G., and Pin, J.
      (1973) Rapport d' Experimentation sur les effets
      du bang sonique. VI, Lyon, Ecole Nationale Veterinaire.
10.   Cottereau, Ph., Place, Ch., and Bastien, G. (1973):
      Rapport d'Experimentation sur les effets du bang
      sonique, III. Lyon, Ecole Nationale Veterinaire.
11.   Davis, P. (1967): British Birds 60, 370-371. Ravens
      response to sonic bang (Corvus corax).
12.   Ellis, N.D., Rushwald, J.B., and Ribner, H.S. (1975).
      J. Sound. Vib. 40, 41.
13.   Emu, F. and Petersen, W.E. (1941), J. Dairy Sci. 24,
      211.
14.   Espmark, 1971, Deer behaviour reactions of reindeer
      exposed to sonic boom.

15.   Espmark, Y., Falt, L., and Falt, B. (1974) Vet. Rec.
      94, 106.
16.   Fausst (1966) Fausst Working Paper 435. Report of
      Veterinary Officer of Ministry of Agriculture.   Ab-
      stracted in an Annotated Bibliography on Animal Re-
      sponse to Sonic Booms and other Loud Noises (1970).
      National Academy of Sciences, National Research Council
      Washington, D.C., U.S.A.
17.   Hawkins, A.D., and Chapman, C.J. 1971. Personal com-
      munication (Marine Laboratory Aberdeen).
18.   Heinemann, J.M. 1969. Proceedings of the Symposium
      on Extra-Auditory Effects of Audible Sound, Annual
      Meeting of AAAS, Boston, MA, December 29, 1969.   Ef-
      fects of sonic booms on the hatchability of chicken
      eggs, and other studies of aircraft-generated noise
      effects on animals.
19.   Hoffman, H.W., Marsh, R.R., and Stein, N. (1969) J.
      Comp. Physiol. Psychol. 68:280.
20.   Hubbard, H.H. 1968. Physics Today, 21,31, Sonic booms.
21.   Rawlins. Personal communication and Fausst Working
      Paper 435, 1966, Effects of sonic booms on animals
      during Exercise Westminster (Report of Veterinary
      Officers of the Ministry of Agriculture, Fisheries
      and Food).
22.   Report 1973. Report of 2nd Meeting of International
      Civil Aviation Organization Sonic Boom Committee,
      Montreal, Canada.
23.   Robertson, W.B., Jr. 1970. Proceedings of the 15th
      International Ornithological Congress, The Hague,
      Holland, August 30-September 5, 1970. Mass hatching
      failure of Dry Tortugas Sooty Terms.
24.   Ruddlesden, F. (1971), Technical report 71084. Royal
      Aircraft Establishment (Ministry of Defence), Farn-
      borough, Hants, U.K.
25.   Teer, J.G. and Truett, J.C. (1973), Report No. FAA-RD-
      73-148. Department of Transportation, Federal Avia-
      tion Administration, Washington, D.C., U.S.A.
26.   Thackray, R.I. (1972), J. Sound Vibration 20:519.
27.   Warfield, D. (1973) In Gay, W.I.(ed.) Methods of
      Animals Experimentation, Vol 4, New York, Academic.

COMPREHENSIVE REVIEW ARTICLES / BIBLIOGRAPHIES

1. Bell, W.B. (1972): Animal responses to sonic booms.
    J. Acoust. Soc. Am. 758-765.
2. Bond, J. (1971): Noise: its effect on the physiology
    and behavior of animals. Agric. Sci. Rev. 9, (4), 1-10.
3. Cottereau, Ph. (1972): Les incidences du "bang" des
    avions supersoniques sur les productions et la vie
    animale. Revue Med. Vet. 123:1367-1409.
4. Ewbank, R. The effects of sonic booms on farm animals
    from Gnensell CSG and HILL, F.W.S. (eds.) 1977. The
    veterinary annual 17th issue Bristol:Wright.
5. National Academy of Sciences/ National Research
    Council (1970). An annotated bibliography on Animal
    Response to Sonic Booms and Other Loud Noises.
    Washington, D.C., U.S.A.
6. U.S. Environmental Protection Agency (1971) Effects
    of Noise on Wildlife and Other Animals. Washington,
    D.C., U.S.A., Report NTID 300.5.
7. Sonic boom exposure effects: Report from a Workshop:
    Editor R. Rylander, Stockholm. 1972. Journal of
    Sound and Vibration, 1972, 20(4):477-544.

WILDLIFE AND ELECTRIC POWER TRANSMISSION

David H. Ellis

Institute for Raptor Studies
Tucson, Arizona

John G. Goodwin, Jr.

University of Arizona
Tucson, Arizona

James R. Hunt

Salt River Project
Phoenix, Arizona

INTRODUCTION

In this presentation we emphasize three related topics: first, we briefly introduce the major areas of wildlife-powerline interactions; second, we present data on the degree of wildlife utilization of powerline corridors and power towers; and finally, we discuss the physical environment on and around high voltage power towers. We have attempted to minimize duplication of coverage with the companion paper in this volume (Lee and Griffith 1977).

GENERAL AREAS OF WILDLIFE POWERLINE INTERACTION

The following general areas of wildlife powerline inter-action deserve attention: (1) animal displacement during the construction and maintenance of powerlines and corridors, (2) wildlife avoidance of powerlines and corridors, (3) direct wildlife mortality due to lines and support structures, (4) possible negative effects of line corona and electrochemical oxidants on wildlife, (5) wildlife benefits due to the presence of transmission support towers, and (6) negative effects of wildlife on transmission systems.

The first consideration (biological changes associated with constructing and maintaining a powerline corridor) has been more completely studied than any of other questions. Numerous "environmental impact" studies are available which treat the effects on plants and animals of opening a swath through the forest, desert, or whatever. Likewise the gross physical impacts associated with powerline construction (erosion, steam siltation, etc.) have been identified (e.g., Kitchings et al. 1974).

The second consideration, wildlife avoidance of powerline corridors, is an important area that has been little treated. Klein (1971) alleged that reindeer (Rangifer tarandus) are deterred from crossing corridors because of corona noise, but we know of no published studies on the response of such wilderness species as the grizzly bear (Ursus horribilis) to powerlines. Gulls (Laridae) (Gunter 1956), and ducks and geese (Anseriformes) (Fog 1970) at least sometimes avoid powerlines. Even if corona noise proves unimportant in causing wildlife to avoid the corridors, the added noise associated with maintenance surveys may result in temporary or permanent displacement.

Third, certain kinds of powerline caused wildlife mortality are well documented. Many species of birds die in collision with tall radio and TV towers, airport ceilometers, powerlines and their support towers, and other man-made obstacles (Willard and Willard unpubl. ms.). Collisions with powerlines and their support structures have been documented for many species including: the trumpeter swan (Olor buccinator)(Banko 1960), mute swan (Cygnus olor) (Harrison 1963), ducks and geese of several species (Mendall 1958, Cornwell and Hochbaum 1971, Stout and Cornwell 1976), sandhill cranes (Grus canadensis) (Walkinshaw 1956), the white stork (Ciconia ciconia) (Jarvis 1974), sage grouse (Centrocercus urophasianus)(Bordell 1939), and even a hummingbird (Selasphorus sasin) (Hendrickson 1949). Powerline collisions generally involve few birds but can be regarded as serious mortality factors in some cases. In the endangered Japanese crane (Grus japonesis) about 12 percent (ca 24 birds) of the annual mortality due to wire strikes (Archibald pers. comm.). The eagle owl (Bubo bubo) has reportedly been largely eliminated from most of its former range in Switzerland as the result of wire strikes (Herren 1969).

Key locations where birds are especially vulnerable (e.g., mountain passes along important migration corridors) should be avoided in siting powerlines. Problems areas should be identified and corrected.   This is one research area where more data are needed even before assessing the seriousness of the problem.

Electrocution is a significant cause of eagle mortality in the western United States (Dickinson 1957; Miller et al. 1975) and cape vulture (Gyps caprotheres) mortality in southern Africa (Marcus 1972; Jarvis 1974).   Other birds have suffered to a lesser extent (e.g., Diler 1954).   Due to the wider spacing of conductors, powerlines greater than 88 kV have not been implicated as causing electrocutions.   Most lines for local power distribution, however, are less than 88 kV and many pole configurations cause electrocutions.   Utility companies have cooperatively financed the design of pole modifications and new conductor configurations which are less dangerous to birds (Miller et al. 1975).

The power company financed recommendations suggest that only 2 percent of existing power poles (preferred raptor perches) need to be modified (Anderson 1975), but we lack detailed studies on: (1) the extent of the electrocution problem, (2) the degree of change necessary to eliminate the problem, and (3) the degree of success in eliminating the problem.   Reason suggests that, even if the power companies finance this research, the field work should be performed by biologists with no vested interest in the outcome.   Laws should also require power companies to report all dead birds encountered along their lines.

Another mortality factor is the legal and illegal hunting of birds and mammals along powerline rights-of-way. Hunting pressure on game mammals is treated in greater detail later.   In an effort to quantify the number of eagles and hawks illegally shot, Ellis et al.(1969) recovered 30 dead raptors in one day from a 19 km stretch of utility poles paralleled by a gravel road in western Utah (Table 1-following page).   Most of the birds were long dead, but those autopsied showed shot or bullet holes.   Public access to maintenance roads should be limited in problem areas.   Power companies should be responsible for identifying these areas.

The fourth general environmental consideration dealt with possible harmful effects associated with line corona and electrochemical oxidants.   Among the electrochemical oxidants produced on extra high voltage (EHV) and ultra high voltage (UHV) lines, ozone is probably the only toxin produced in high enough levels that it may present a problem to wildlife.   Many plants are more sensitive to ozone than are animals (Miller et al. 1975), but ozone concentrations at ground level have not

Table 1

Raptorial Birds Found Dead on the Initial Check
(April 29, 1967) of a 19 km Survey Route in Western Utah

Common Name	Scientific Names	Number Found
Golden Eagle	Aquila chrysaetos	14
Rough-legged hawk	Buteo lagopus	8
Swainson's hawk	Buteo swainsoni	1
American kestrel	Falco sparverius	1
Unidentified hawk[1]	Buteo spp.	5
Great horned owl	Bubo virginianus	1
Total		30

[1] These birds were identifiable only to genus owing to advanced state of decomposition.

been significantly altered even along 765 kV lines (Frydman et al. 1972; Scherer et al. 1972; Fern and Brabets 1973). Transmission towers are important perches and nesting sites for significant numbers of raptors and other birds. Field and laboratory studies are needed to determine if line-produced ozone has negative effects on these birds. In a superficial check, two red-tailed hawks (Buteo jamaicensis) were removed at 4 weeks of age from a 500 kV line in Oregon, reared to fledging in captivity, and released. Neither bird empirically showed signs of impairment.

Noise produced by low voltage transmission lines is probably not a significant health hazard to nesting birds, but EHV lines produce noise levels that may cause some terrestrial species to avoid transmission corridors. Although it is improbable that cochlear damage will result in animals on the ground, tower nesting birds should be studied to determine if they suffer hearing impairment. All previously published studies have emphasized noise levels, oxidant levels, etc. at the ground, but data on the acoustic environment on the tower are presented later in this paper.

There is an accumulating body of literature suggesting that vertebrates are able to detect and navigate by the earth's magnetic field (Walcott 1974; Rommel and McCleave 1973). Several groups of fish are also capable of communicating via weak electrical discharges (Hopkins 1974). The effects of introducing EHV and UHV transmission lines into the environment of creatures dependent on either the earth's magnetic field for navigation or on self-generated electric impulses for communication are entirely unknown.

The fifth consideration dealt with wildlife benefits. General observations from various biologists in the northwest United States suggest that deer (Odocoileus hemionus and O. virginianus), elk (Cervus canadensis), and bighorn sheep (Ovis canadensis) probably benefit from the additional food plants along powerline corridors. Goodwin's (1975) quantitative study, indicating that the above species do use the corridors as heavily as other woodland clearings, will be discussed in detail hereafter. As discussed below in greater detail, ravens (Corvus corax), eagles and hawks (Accipitridae), and occasionally other birds locally use power towers for nest sites. Wild turkeys (Meleagris gallopavo) heavily utilize power towers and even the conductors as roost sites in west Texas (H. Kothman, 1971). Cleared powerline corridors may provide additional wildlife feeding areas and power towers may provide supplemental nest sites, roosts, foraging perches, and congregation areas (Schmidt 1973), but

these benefits may be offset if migratory birds (or mammals)
are encouraged by the cleared powerline corridor to grossly
alter their traditional migration routes as has been reported
by Carothers and Johnson (1973).

The sixth consideration concerns the negative effects of
wildlife on powerlines. There are well documented accounts of
woodpecker (Picidae) damaging utility poles (Willard and
Willard unpubl. ms.). There are also accounts both of raptor
nests causing flashovers (Nelson and Nelson 1976) and excrement from large birds causing outages. An investigation to
determine the cause of 32 unexplained outages in 25 months
along 500 kV lines in the northwestern United States led to
the implication of eagles and possibly a few other species
as most probable cause. Supposedly the eagle, either will-
fully or by accident, voids near the insulator string and
the electrolytic fluid allows the current to arc between the
conductor and the ground wire or tower (West et al. 1971).

In solving the raptor nest problem, the stick matrix can
be trimmed to prevent flashovers (Anderson 1975) or moved to
a more desirable position and secured to the tower (Hoaglin
unpubl. ms.). The excrement problem can be reduced by placing
bird guards above the top of the insulator string (West et al.
1971).

UTILIZATION OF POWERLINE CORRIDORS BY LARGE GAME MAMMALS IN WESTERN NORTH AMERICA

Several of the factors described above could reasonably
influence big game use of and movement across transmission
line corridors. When elk are prevented from completing their
migration, severe range overuse and deterioration may occur
(Craighead et al. 1972:11). Mace (1971:9) found that hunting
pressure could delay elk migration, and Ward et al. (1973:337)
reported that an interstate highway acted as a barrier to
elk movement. However, thousands of elk in the Jackson Hole
herd annually cross highways during their migration (Anderson
1958). Scandinavian reindeer herders claim " that reindeer
are disturbed by the newly constructed power lines, which are
foreign objects in otherwise familiar surroundings. For this
reason, reindeer are afraid to travel beneath the power lines,
particularly during the first year or two after their re-
construction." (Klein 1971:396). In contrast, biologists
in the northwestern United States believed that transmission
line rights-of-way were beneficial to elk (C. c. nelsoni and
roosevelti), deer, black bear (Ursus americanus), and bighorn sheep because of added forage production in the cleared
corridor (Harper 1971, Taber et al. 1972, Brown pers. comm.

1974, Blackburn pers. comm. 1974).

To gather quantitative data on the magnitude of big game
utilization of an EHV transmission corridor, surveys were
conducted from July 1974 to June 1975 along the Dworshak-Hot
Springs 500 kV line which crossed both winter and summer
ranges of several big game species (Goodwin 1975). The study
was designed to determine if the transmission corridor,
cleared through the moist deciduous forests of northern Idaho
and western Montana was used more or less than nearby forest
clearings. Five sites, each consisting of a segment of right-
of-way and a nearby clearcut or natural opening, were studied
to compare intensity of animal use. The control clearings
were 100-400 m from the powerline right-of-way, and each
control area had slope, aspect, and elevation similar to the
powerline area with which it was compared. However, the controls differed somewhat from the powerline rights-of-way in
size, shape, and vegetative cover (Table 2-following page)

Animal observations were made in three ways. Direct observations during the early morning and late evening were
made from vantage points allowing simultaneous viewing of a
right-of-way and control pair. Time lapse super 8 mm cameras,
exposing one frame every 30 seconds, were positioned to sup-
plement the direct observations of study areas during daylight
hours. Track counts were made on parallel, one meter wide,
bare earth transects through each right-of-way and control
area to document both diurnal and nocturnal animal movements.
After snowfall initiated the fall migration from summer to
winter ranges, additional transects were conducted para-
lleling the power corridor at ca 400 m distance. Along these
transects the transmission line was generally out of sight
and inaudible. When animal tracks were encountered, they
were followed as they approached and either crossed or re-
treated from the right-of-way. The animal's direction and
rate of travel gave some insight as to its behavior as it
approached and departed from the powerline corridor.

Vegetative analyses including percent ground cover of
grasses, forbs, and shrubs less than 1.5 m in height were
determined at each site using Daubenmire's (1959) method.
Three transects were run in each right-of-way and control
area: one down the center, one in the clearing 9 m from
the forest edge, and one in the forest 9 m from the edge.
Each transect consisted of 40 plots, each 0.10 m ^{2} in
size, spaced at 1.5 m intervals. The percent of ground
covered by grasses, forbs, and shrubs was classed as 0-5%,
5-25%, 25-50%, 50-75%, 75-95%, or 95-100%. The mid-points
of these categories were used to compute average ground cover.

Table 2

Physical and Vegetative Factors for Study Areas

Study Area	Size (ha.)	Length of Track Count Transect (m)	Down Slope	Aspect	Elevation (m)	Percent Ground Cover		
						Grass	Shrubs	Forbs
#1 Right-of-Way	3.6	460	15	N	1220	17	19	13
#1 Control	10.1	460	25	N	1220	5	24	32
#2 Right-of-Way	0.9	185	30	E	1440	24	11	16
#2 Control	1.8	185	30	E	1400	11	21	28
#3 Right-of-Way	1.0	160	30	SW	795	47	33	6
#3 Control	1.1	160	35	SW	795	21	15	4
#4 Right-of-Way	0.9	115	35	SW	975	18	31	9
#4 Control	1.1	115	35	SW	975	34	27	18
#5 Right-of-Way	1.3	185	10	S	885	28	26	6
#5 Control	1.3	185	15	S	885	28	5	27

The data on animal utilization of study areas 1 and 2
from August to December were very inclusive. During 310 observation hours only 4 deer, 6 elk, and 2 bear were seen.
Thirty-five rolls of super 8 mm film covering 119 days showed
only another 5 elk, 2 deer, and 2 bear crossing the right-
of-way. Track count data (Table 3-following page) also indicated very low levels of animal activity. Yet the forest
bordering the study areas was known to support relatively high
densities of deer, elk, and bear based on sightings, tracks
crossing forest roads, fresh droppings, and bugling elk.
Northern Idaho experienced unusually warm, dry weather, during
this period. This may have caused an unfavorable temperature
regime for big game species in the right-of-way and control
clearings.

When snowfall initiated the fall migration in late November, animal use of the controls and right-of-way clearings
increased. During the first two weeks of December, 9 sets of
deer tracks and 87 sets of elk tracks were followed as they
approached the right-of-way. All but two of the elk maintained a steady direction and rate of travel toward and across
the right-of-way. Two elk approached within 25 m of the corridor and began feeding on brush, then moved away from the
right-of-way without crossing. There appeared to be no hesi-
tation by these animals to enter or cross the corridor.

Data were collected on study areas 3,4, and 5 from March
to May 1975. This is normally a critical time for big game
species because of deep snow and low food supplies. Snow in
the open right-of-way and control areas melted faster than
that under the forest canopy, and the clearings contained a
higher density of palatable plant species. Consequently, big
game use of all forest clearings was very intense.

Track counts showed deer activity average 23 tracks/km/
day, and elk use average 5 tracks/km/day. This contrasts with
the fall activity levels of 1.6 deer tracks/km/day and 0.5 elk
tracks/km/day. To compare the track count data for the right-
of-way and control pairs a Chi-square test of difference was
used. Expected values were adjusted for each site according
to the percent ground cover values from Table 3. There was no
significant difference in deer use of the three areas. Elk
use was significantly higher on one right-of-way site than on
the associated control (Table 4-following Table 3).

Time lapse camera data supported the results from the
track counts. From 21 rolls of film (covering 63 days) 264
deer and 18 elk were recorded on the study areas for 8,696
and 503.5 minutes respectively. When the numbers of animals
seen in the right-of-way and control areas were weighted according to the amount of available vegetation, there was

Table 3

Track Count Data for Deer and Elk On and Off the Right-of-Way Of a 500 kV
Transmission Line in Northern Idaho

Area	Dates Included In Count	Total Days	Deer Tracks[1] Right-of-Way	Deer Tracks[1] Control	Elk Tracks[1] Right-of-Way	Elk Tracks[1] Control
1	8/28 - 10/28	56	28	31	3	4
2	10/28 - 11/27	18	7	11	13	6
3	3/19 - 4/9	21	234	121	21	4
4	3/31 - 4/24	24	206	152	22	45
5	4/9 - 5/9	30	46	32	60	17

[1]Track counts represent the number of times the transect was crossed rather than an actual number of animals.

Table 4

Chi-Square Values for Track Counts on Right-of-way and Control Pairs

Area	Percent Ground Cover	Tracks Observed	Tracks Expected[1]	O-E	$(O-E)^2$	$(O-E)^2/E$
		Area 3 Deer Tracks				
Right-of-Way	86	234	241.5	7.5	56.25	0.23
Control	40	121	113.5	7.5	56.25	0.50
						$X^2 = 0.73$ [2]
		Area 3 Elk Tracks				
Right-of-Way	86	21	17	4	16	0.94
Control	40	4	8	4	16	2.00
						$X^2 = 2.94$ [2]
		Area 4 Deer Tracks				
Right-of-Way	58	106	108.5	2.5	6.25	0.06
Control	79	152	149.5	2.5	6.25	0.04
						$X^2 = 0.10$ [2]
		Area 4 Elk Tracks				
Right-of-Way	58	22	28.1	6.1	37.21	1.32
Control	79	45	38.9	6.1	37.21	0.96
						$X^2 = 2.28$ [2]
		Area 5 Deer Tracks				
Right-of-Way	60	46	39	7	49	1.26
Control	60	32	39	7	49	1.26
						$X^2 = 2.52$ [2]
		Area 5 Elk Tracks				
Right-of-Way	60	60	38.5	21.5	462.25	12.01
Control	60	17	38.5	21.5	462.25	12.01
						$X^2 = 24.02$ [3]

[1] Adjusted according to the percent ground cover of grasses, forbs, and shrubs by the following formula:

$$\text{Tracks Expected On Plot} = \frac{\text{\% Ground Cover (PGC) of Plot}}{\text{PGC of Plot + PGC of Paired Plot}} \times (\text{Tracks Observed (TO) on Plot + TO on Paired Plot})$$

[2] Not significantly different at $P < 0.05$.

[3] Significantly different at $P < 0.05$.

no significant difference between animal use of the areas.

Deer were filmed in the study areas for periods as short as 30 seconds (1 frame) and as long as 262 minutes. Elk use varied between 3 minutes and 77 minutes. Deer remained in the right-of-way for an average of 34.3 minutes and in the control for 31.1 minutes. Corresponding values for elk were 28.4 minutes in the right-of-way and 27.7 minutes in the control Both deer and elk spent most of this time feeding.

In summary, deer and elk used the clearings primarily for feeding, and the intensity of use was proportional to the abundance of food. The average length of visit was similar for the rights-of-way and the controls, thereby indicating that the transmission line did not make the right-of-way less attractive.

The most significant influence of the transmission corridor on elk movement was hunter activity on the cleared right-of-way and access roads. Although segments of the right-of-way were closed by locked gates, some hunters broke the gates in order to drive onto the right-of-way. Others went around the gates on motorcycles. Many hunters parked at the gates and walked to the right-of-way.

During the general elk and deer hunting season, 5 October to 10 November, 268 people were seen on foot actively hunting in the right-of-way, control areas, or along the access roads. Many others simply drove the roads hoping to see game. Most of this pressure was concentrated in the first and last weeks of the season.

Of 107 hunters interviewed during the study, 98 said they were hunting in one of the clearings or along the roads. Only 9 people indicated that they had hunted more than 0.8 km from a road or clearing. Studies in eastern United States also indicated that hunters concentrated along roads and trails (Gramlich 1965, James et al. 1964a, 1964b). Where transmission line access roads are built in a forest with few roads, the additional access may aid in the hunting of under-harvested herds. The increased harvest could result in better herd survival during hard winters, but in areas already heavily harvested, new access roads could stress local herds.

RAPTOR UTILIZATION OF POWER TOWERS AS PERCHES AND NEST SUPPORT STRUCTURES

Raptors of many species are known to utilize power towers as nest support structures. The Idaho Power Company identified 32 raptor or raven nests along its lines in 1973 (Nelson and Nelson 1976). Red-tailed hawks and ravens make use of the power towers in many western states, and along some treeless stretches in western Oregon and Washington, there is an average of one tower nest every 8-12 km. Martial eagles (<u>Polemaetus</u> <u>bellicosus</u>) infrequently nest on metal power towers in South Africa (Dean 1975).   In a study initiated in 1973 by the Public Service Company of New Mexico, A.B. Rodney (pers. comm.) documented heavy use of power towers as perches by seven species of raptors in various vegetation types in New Mexico.

Some power company financed studies assert that, for thermal reasons, tower nests are superior to south facing cliff nests (Anderson 1975; and Nelson and Nelson 1976). In the only study now available comparing productivity, D. S. Gilmer and J. M. Wiehe (unpubl. ms.) found that ferruginous hawks (<u>Buteo</u> <u>regalis</u>) in North Dakota fledged fewer young from power tower nests than were fledged from nearby non-tower sites eventhough the average clutch size was slightly higher for tower nests than elsewhere. The advantages incurred by nesting on well-ventilated towers may be offset by the dangers associated with excess exposure to the sun and wind.   Six of 8 nests reported by Idaho Power Company maintenance crews in November 1973 (Nelson and Nelson 1976) and visited by us in April 1974 were on the ground below the pole or tower.   One of these held 2 freshly broken golden eagle (<u>Aquila</u> <u>chrysae-</u> <u>tos</u>) eggs.

In an effort to document the degree of raptor utilization of power towers as nest sites, we began checking for heavy use areas in 1974 and continued sporadically until the present time. The greatest concentration of ferruginous hawk nests found by Gilmer and Wiehe (unpubl. ms.) was 5 nests in 16 km. The closest nests were 1.3 km distant. The area of greatest concentration encountered during our study was along a 230 kV line in northern Sonora, Mexico. The area consists of flat and rolling terrain heavily invaded by brush. The major brushy overstory species are, in order of importance: mesquite (<u>Prosopis</u> <u>juliflora</u>), palo verde (<u>Cercicium</u> spp.), creosote (<u>Larrea</u> sp.), and ironwood (<u>Olneya</u> <u>tesota</u>). On an initial survey of the line in August 1976, 39 nests were located in 83 km. In May 1977 the line was again visited to determine occupancy of the nests. Along a 141 km survey route, 65 nests were located of which 17 were inactive, 23 were active red-tailed hawks nests, 24 were active raven nests, and in 1 a Harris hawk (<u>Parabuteo</u> <u>unicinctus</u>) was incubating. In a 105 km zone of concentration there were 19 red-tailed hawk nests (5.5 km/nest), 24 raven nests (4.4 km/nest), and 1 Harris' hawk nest for a total of 44 active nests (2.4 km/nest). In one segment of this zone of concentration 5 adjacent towers contained nests (d=1.5 km): the first and fifth were active red-tailed hawk nests, and the remaining three were active raven nests.

In the Sonoran study area, nests were placed at one of two locations on the metal power towers (Fig. 1) either at point A (immediately above the insulator strings) or below the phases at point B.

Fig. 1 - Red-tailed hawk nest on 230 kV ac tower in Sonora, Mexico (photographed by Greg Depner). This nest is at the most commonly used tower location (A). The alternate location (B) is also identified.

Of 24 raven nests 20 were at point A. Only 1 of 22 red-tail
nests considered in the analysis was at point B. The locations of the nests on the towers are important in the next
section when describing the physical environment available
to the birds.

THE PHYSICAL ENVIRONMENT NEAR EHV AND UHV TRANSMISSION LINES

In this section we attempt to quantify some of the physical factors surrounding EHV and UHV power towers. Hopefully
subsequent studies will determine if the levels reported for
audible noise, electric and magnetic field strengths, etc.
are disadvantageous for each particular species.

In an effort to describe the physical environment available to mammals on the ground, noise levels, electric field
strength, and magnetic field strength were recorded along the
500 kV study line in Idaho. Three types of conductor bundles
were in use in the study area: the single, double, and
triple-conductor forms. The single conductor was the loudest.
Audible noise was measured in the study areas with a General
Radio type 1551-C sound level meter capable of discerning
levels between 30-140 decibels (dB). Measurements were made
at ground level using the "A" scale whenever animals were
observed in the study area and under varying weather conditions to determine normal and extreme dB (A) levels. Typical
ground readings along the single conductor phase line varied
from 45 to 55 dB (A). During the rainy weather readings of
62-68 dB (A) were common. The two conductor bundle normally
produced 35-40 dB (A) readings, and levels below the 30 dB
(A) threshold of the sound meter were not unusual. Rain produced noise levels from 45-56 dB (A). Sound from the triple-
conductor bundle was generally below 30 dB (A), however, a
reading of 52 dB (A) was recorded during a light shower.

Readings made in the control areas away from the
corridor were generally below 30 dB (A). One reading of
44 dB (A) was taken in the control for Area 2 during light
rain.

During several days of constant 55-60 dB (A) noise
levels, track counts indicated that the normal movement patterns of both deer and elk were unaltered. Two elk were known
to have crossed the right-of-way during a snowstorm when noise
levels were about 63 dB (A). Also, a family of coyotes (Canis
latrans) was observed feeding and playing directly under the
conductors when noise was 63 dB (A). Ravens, blue grouse
(Dendragapus obscurus), ruffed grouse (Bonasa umbellus),

several species of hawks, and numberous songbirds were seen
sitting under or flying near the conductors when noise levels
exceeded 60 dB (A). A herd of 59 bighorn sheep was observed
feeding and bedded in the right-of-way during a light rain
when the noise level was 53 dB(A). From these observations
it appears that many wildlife species are not disturbed by
transmission line audible noise of up to 60 dB (A). The
temporal pattern of noise is as important as the volume. A
person breaking a stick or clapping his hands may frighten
nearby deer or elk, when a relatively constant noise of the
same volume would produce no response in the same animals.
To the human ear, transmission lines produce a steady noise
much like the sound of leaves rustling in the wind or water
flowing in a stream. Although the volume may vary greatly
with time, it seldom changes abruptly.

Electric field strengths were measured under field conditions with a meter constructed by the Bonneville Power Administration (EPA) and compared with values predicted by a
BPA computer program based on voltage records kept at BPA
substations. The highest measured value was 7 km/m beneath
an outer conductor with a ground clearance of 10.6 m. Field
strengths under the outer conductor commonly measured 3-4 kV/
m where ground clearance was 11-20 m. Readings dropped rapidly as one moved away from the conductors and generally were
below 2 kV/m at the forest edge.

Electric fields of the strength found beneath 500 kV
transmission lines are generally not perceptible to humans.
In one study the threshold for human perception (hair, stimulation, tingling sensation) for more than 10 percent of the
cases was between 10 and 15 kV/m. Less is known about wildlife perception of electric fields. Deno and Zafanella (1975:
263-264) also reported no visible changes in grazing, feeding,
and drinking habits of cattle on damp ground in electric
fields up to 18 kV/m. However, they added that on dry ground,
spark discharges might occur between grounded objects and
bodies of animals. In this study and along other 500 kV transmission lines rights-of-way we failed to observe wild or
domestic animals exhibit any reluctance to enter the field
area.

In the Americas the geomagnetic field at the earth's
surface ranges from 0.25 gauss in southern Brazil to 0.60
gauss in northern Canada (Hess 1968). Magnetic field
strengths due to EHV and UHV transmission lines are generally
of the same order of magnitude as the geomagnetic field but
are greater within a radius of a few meters of the conductors.

Since the magnetic field strength is inversely proportional to distance from the conductor, magnetic fields at the edge of the right-of-way are generally well below the geomagnetic field strength (Deno 1976).

Magnetic field strengths were calculated from substation records of voltage and current loads for the Dworshak-Hot Springs line. The maximum computed strength was approximately 0.10 Gauss at ground level. As with electric fields, magnetic fields have had biological effects (Barnothy 1964), however, the field strengths resulting in biological changes were usually many times greater than those recorded beneath the transmission lines.

Computed theoretic values from transmission line magnetic field strength are presented in Table 5(following page). Since the magnetic field is a function of the line loading (amperage), great variations will occur for a given line due to time-of-day, season, emergency conditions, etc., in addition to variations between lines due to differences in conductor type and tower geometery. The calculations of Table 5 were made using the "typical" tower dimensions in Table 6 (following Table 5) and assumed line currents which represent extreme line loading. Because magnetic field strengths may vary rapidly and over a wide range, the calculations give only an order-of-magnitude approximation of magnetic field strength for any moment in time.

The physical environment available to tower nesting birds has been less carefully documented than that for terrestrial life. The studies cited throughout this report emphasize ground level values for line corona, force fields, etc. There has been some consideration given to effects of electric and magnetic fields near the conductors due to concern over these effects on line men who engage in live line maintenance (e.g., work in the close vicinity of energized conductors) (Singewald 1973). However, because live line work is done in fair weather when corona noise levels are low, the maximum audible noise levels encountered were well below the levels considered hazardous to humans. As a result, there are no published data known to us for corona noise at or near the conductors. In an effort to fill this void we calculated the theoretical values from the semi-empirical formulas published by the Electric Power Research Institute (1975). In interpreting Table 6 it is important to emphasize that the audible noise values are approximations for a hypothetical average tower-conductor configuration and under extreme weather conditions (i.e., good weather values will be

Table 5

Magnetic Field Strengths (B)
On and Near Typical EHV and UHV Power Towers

Voltage, kV	Current amperes/ phase	B at 0.5 m above center conductor, gauss	B at closest point on tower, gauss[1]	B at mid tower, gauss[1]
345	1200	4.8	0.58	0.23
500	2000	8.0	0.73	0.29
765	3000	12.0	0.80	0.37
1100	4000	16.0	0.96	0.39

[1]Calculations are based on a flat phase spacing design as in Fig. 1. Distances are in Table 6.

Table 6

HEAVY RAIN CORONA NOISE LEVELS
ON AND NEAR TYPICAL EHV AND UHV POWER TOWERS[1,2]

Voltage, kV	Number subconductors	Bundle diameter, cm	Subconductor diameter, cm	Phase spacing, m	Voltage gradient at center conductor surface, kV/cm	Voltage gradient at outer conductor surface, kV/cm	Audible Noise[3] 4 m above center conductor, dB(A)	Distance to nearest point on tower, m	Audible noise[3] at nearest point on tower, dB(A)	Distance to mid-tower, m	Audible noise[3] at center of tower, dB(A)	Minimum conductor-ground clearance, m	Audible noise at ground, dB(A)
345	1		4.6	7.5	14.3	15.2	76	3	69	9	65	9	61
345	2	45.7	2.8	7.5	15.8	17.1	72	3	65	9	61	9	57
500	1		6.35	10	15.4	16.2	84	4	76	12	71	11	69
500	2	45.7	4.6	10	14.5	15.5	77	4	69	12	66	11	62
500	3	52.8	3.3	10	15.0	16.2	71	4	66	12	62	11	58
765	4	64.7	3.6	14	16.6	17.9	81	5.5	72	14	69	14	65
1100	8	101	4.3	18.5	12.4	13.5	80	6.5	69	18	65	19	60

[1] Calculations are based on a flat phase spacing design as in Fig. 1.

[2] For distances between conductors and various points on the towers, see Electric Power Research Institute (1975).

[3] Audible noise calculations at tower were made using voltage gradient values 11% higher than the tabulated average values as recommended in Fig. 6.6.2 of Electric Power Research Institute (1975).

much lower and will vary considerably from the values in the table according to the actual voltage of the line, spatial orientation of the bundle, horizontal or vertical, symmetry of subconductors in the bundle, subconductor size, integrity of the conductor surface, tower configuration, etc.).

In closing this section we emphasize that, even though the theoretical and reported values for corona noise and other elements of the artificial powerline environment are of the order of magnitude that they may be harmful to wildlife, there are no studies known to us where the types of animals which will live in close association with transmission lines have been experimentally subjected to these various synthetic environmental factors at the levels they will likely encounter in the wild.

CONCLUSIONS

Hundreds of thousands of miles of transmission lines have been introduced into our natural environment. These lines and their corridors can be damaging or beneficial to wildlife communities depending on how they are designed, where they are placed, and when they are constructed and maintained. With the current trend toward UHV systems, new problems (associated with additional increments in audible noise, electric and magnetic force fields, etc.) must be addressed. We recommend the following areas for careful study: (1) the response of wilderness species to transmission lines and line construction and maintenance activities (2) the magnitude of bird collision and electrocution mortality, (3) the response of power corridor and power tower inhabiting wildlife to laboratory and field doses of electrochemical oxidants, corona noise, electric and magnetic fields, etc., (4) the productivity of tower inhabiting birds compared with nearby non-tower nesters, and (5) the influence of powerline corridors on mammalian and avian migration patterns. It is our hope that the questions identified in this study will help stimulate further research so that we can maximize wildlife benefits and minimize wildlife detriments.

REFERENCES CITED

1.   Anderson, C. C. 1958. The elk of Jackson Hole. A re-
     view of Jackson Hole elk studies. Bull. 10. Wyoming
     Game and Fish Comm., Cheyenne. 184 p.
2.   Anderson, W. W. 1975. Pole changes keep eagles fly-
     ing. Transmission and Distribution (Nov):28-31.
3.   Banko, W. E. 1960. The trumpeter swan, its history,
     habits and population in the United States. North
     American Fauna, No. 63. 214 pp.
4.   Barnothy, M. F., ed. 1964. Biological effects of
     magnetic fields. Plenum Press, New York. 2 vols.
5.   Borell, A. E. 1939. Telephone wires fatal to sage
     grouse. Condor 41:85.
6.   Carothers, S. W., and R. R. Johnson. 1975. Water
     management practices and their effects on nongame birds
     in range habitats. Pages 210-222 in Proc. of the
     symposium on management of forest and range habitats
     for nongame birds. USDA For. Ser. Gen. Tech. Rep.
     WO-1.
7.   Cornwell, G., and H. A. Hochbaum. 1971. Collisions
     with wires--a source of anatid mortality. Wilson Bull.
     83:305-306.
8.   Craighead, J. J., G. Atwell, and B. W. O'Gara. 1972.
     Elk migration in and near Yellowstone National Park.
     Wildl. Monogr. 29. 48 pp.
9.   Daubenmire, R. 1959. A canopy-coverage method of
     vegetational analysis. Northw. Sci. 33:43-64.
10. Dean, W. R. J. 1975. Martial eagles nesting on high
     tension pylons. Ostrich 46:116-117.
11. Deno, D. W. 1976. Transmission line fields. IEEE
     Trans. PAS 95:1600-1611.
12. Deno, D. W. and L. E. Zaffanella. 1975. Electrostatic
     effects of overhead transmission lines and stations.
     Pp. 248-280 in Electric Power Research Institute.
     Transmission line reference book 345 kV and above.
     General Electric Co., Palo Alto, Calif.
13. Dickinson, L. E. 1957. Utilities and birds. Audobon
     Mag. 59:54-55, 86-87.
14. Dilger, W. C. 1954. Electrocution of parakeets at
     Agra, India Condor 56:102-103.
15. Electric Power Research Institute. 1975. Transmission
     line reference book 345 kV and above. General Electric
     Co., Palo Alto, Calif. 393 pp.
16. Ellis, D. H., D. G. Smith and J. R. Murphy. 1969.
     Studies on raptor mortality in western Utah. Great
     Basin Nat. 29:165-167.

17. Fern, W. J., and R. I. Brabets. 1973. Field investi-
     gations of ozone adjacent to high voltage transmission
     lines. IEEE paper no. T74057-6.
18. Fitzner, R. E. 1975. Owl mortality on fences and
     utility lines. Raptor Research 9:55-57.
19. Fog, J. 1970. Om andefugle contra elledninger (English
     Summary). Flora og Fauna 76:141-144.
20. Frydman, M., A. Levy, and S. E. Miller. 1972. Oxidant
     measurements in the vicinity of energinzed 765 kV
     lines IEEE paper no. T72551-0.
21. Gilmer, D. S., and J. M. Wiehe. Unpubl. ms. Nesting
     by ferruginous hawks and other raptors on high voltage
     powerline towers.
22. Goodwin, J. G., Jr. 1975. Big game movement near a
     500 kV transmission line in norther Idaho. Bonneville
     Power Administration report. 56 pp.
23. Gramlich, F. J. 1965. A study of factors related to
     low deer harvests in a portion of eastern Maine.
     Unpubl. ms. thesis, Univ. Maine, Orono. 106 pp.
24. Gunter, G. 1956. On the reluctance of gulls to fly
     under objects. Auk 73:131-132.
25. Harper, J. A. 1971. Ecology of Roosevelt elk. Oregon
     State Game Comm. P-R Proj. W-59-R. 44 pp.
26. Harrison, J. 1963. Heavy mortality of mute swans from
     electrocution. Ann. Rep. the Waterfowl Trust - 1961-62
     14:164.
27. Hendrickson, J. R. 1949. A hummingbird casualty.
     Condor 51:103.
28. Herren, H. 1969. Status of the peregrine falcon in
     Switzerland. Pp. 231-238 in Hickey, J. J. ed. Pere-
     grine falcon populations, their biology and decline.
     Univ. of Wis. Press: Madison, Wisc.
29. Hess, W. N. 1968. The radiation belt and magnetosphere
     Blaisdell Publishing Co., Waltham, Mass. 548 pp.
30. Hoaglin, E. D. Unpubl. ms. Bird nest problem: a pos-
     sible solution. Bonneville Power Adminstration report.
31. Hopkins, C. D. 1974. Electric communication in fish.
     Am. Sci. 62:426-437.
32. James, G. A., M. Johnson, and F. B. Barick. 1964a.
     A key to better hunting--forest roads and trails.
     Wildl. N. C. 2 pp.
33. James, G. A., M. Johnson, and F. B. Barick. 1964b.
     Relations between hunter access and deer kill in
     North Carolina. Trans. N. Am. Wildl. Nat. Resour.
     Conf. 29:454-463.
34. Jarvis, M. J. F. 1974. High tension power lines as a
     hazard to larger birds. Ostrich 45:262.

35. Kitchings, J. T., H. H. Shugart, and F. D. Story. 1974. Environmental impacts associated with electric transmission lines. Oak Ridge National Lab. report. Contract W-7405-eng-26. 111 pp.

36. Klein, D. R. 1971. Reaction of reindeer to obstruction and disturbances. Science 173:393-398.

37. Kothmann, H. G. 1971. High-rise roosts. Texas Parks and Wildlife 29:26-29.

38. Lee, J. M., Jr., and D. B.Griffith. 1977. Transmission line audible noise and wildlife. (this volume).

39. Mace, R. U. 1971. Oregon's elk. Oregon State Game Comm. Wildl. Bull. 4. 29 pp.

40. Marcus, M. B. 1972. Mortality of vultures caused by electrocution. Nature 238:238.

41. Mendall, H. L. 1958. The ring-necked duck in the northeast. Univ. of Maine Studies, 2nd series, no. 73. 317 pp.

42. Miller, D., E. L. Boeker, R. S. Thorsell, and R. R. Olendorff. 1975. Suggested practices for raptor protection of powerlines. Raptor Research Foundation publication. 19 pp.

43. Nelson, M. W., and P. Nelson. 1976. Power lines and birds of prey. Idaho Wildl. Rev. 28:3-7.

44. Rommel, S. A. Jr., and J. D. McCleave. 1973. Sensitivity of American eels (Anguilla rostrata) and Atlantic salmon (Salmo salar) to weak electric and magnetic fields. J. Fisheries Res. Board of Canada 30:657-663.

45. Scherer, H. N. Jr., B. J. Ware, and C. H. Shih. 1972. Gaseous effluents due to EHV transmission line corona. IEEE paper no. T72550-2.

46. Schmidt, E. 1973. Okologische auswirkungen von elektrischen leitungen und masten sowie deren accessorien auf die vogel (in German, English Summary). Beitr. Vogelkd. 19:342-362.

47. Singewald, M. L., W. B. Kouwenhoven, and O. R. Langworthy. 1973. Medical follow-up of high voltage lineman in AC electric fields. IEEE Trans. PAS 92:1307-1309.

48. Stout, I. J., and G. W. Cornwell. 1976. Nonhunting mortality of fledged North American waterfowl. J. Wildl. Manage 40:681-693.

49. Smith, D. G., and J. R. Murphy. 1972. Unusual causes of raptor mortality. Raptor Research 6:4-5.

50. Taber, R. D., D. R. M. Scott, C. H. Driver, C. Erickson W. Bradley, J. Schoen, and L. Leschner. 1973. Wildlife response to rights-of-way management. Res. project 63-1013. College of Forest Resources. Univ. Wash., Seattle. 24 pp.

51.   Walcott, C. 1974.   The homing of pigeons.   Am. Sci.
       62:542-552.
52.   Walkinshaw, L. H. 1956.   Sandhill cranes killed by fly-
       ing into power line.   Wilson Bull. 68:325-326.
53.   Ward, A. L., J. J. Cupal, A. L. Lea, C. A. Oakley, and
       R. W. Weeks. 1973.   Elk behavior in relation to cattle
       grazing, forest recreation, and traffic.   Trans. N. Am.
       Wildl. Nat. Resour. Conf. 38:327-337.
54.   West, H. J., J. E. Brown, and A. L. Kinyon. 1971.
       Simulation of EHV transmission line flashovers in-
       itiated by bird excretion.   IEEE transaction paper
       71 TP 145-PWR.
55.   Willard, D. E., and B. J. Willard. Unpubl. ms.
       The interaction between some human obstacles and
       birds.

TRANSMISSION LINE AUDIBLE NOISE AND WILDLIFE

Jack M. Lee, Jr.

Bonneville Power Administration
Portland, Oregon

Dennis B. Griffith

Western Interstate Commission for Higher Education
Boulder, Colorado

INTRODUCTION

Transmission line rights-of-way have been a conspicuous part of the landscape for many years. Until the advent of extra high voltage lines (EHV, above 230 kV) it was often assumed that transmission lines had a net beneficial effect on wildlife due to the increased habitat diversity produced by cleared rights-of-way (Egler 1953, 1957).

When the first EHV transmission lines were constructed in the U.S. during the 1950's the transmission facilities themselves began to be a more noticeable component of the right-of-way environment and new kinds of environmental impacts were identified. These included greater visual impacts due to the larger physical size of the facilities and electric field and corona effects due to the higher operating voltages. Of the latter effects, audible noise (AN) due to corona (air ionization) was found to be a source of annoyance to persons living near EHV lines (Perry 1972). Concerns have also been raised about the possible effects of transmission line AN on wildlife (Klein 1971, Villmo 1972, Martinka 1974, Driscoll 1975, Fletcher 1975, Balda and Johnson 1976, Grue 1977).

Noise is usually defined as unwanted sound. It can also be described as an environmental pollutant which is a waste product generated by various human activities (EPA 1974). In general, wildlife species are exposed to many of

the same environmental noises as man. A recent report included an estimate that as many as 13 million American live
in places where noise from cars, buses, trucks, airplanes,
construction equipment, and electrical devices may be
harming their health (Comptroller General 1977).

Although the hearing ability of wildlife varies greatly
among species and may differ significantly from man, a
knowledge of the effects of noise on wildlife aids in understanding the effects of noise on man and vice versa. The
concept of wildlife as an indicator of environmental quality
has been described by a number of authors (Thomas et al.
1973, Jenkins 1972).

Noise can produce in man and animals such effects as
hearing loss, masking of communications, behavioral changes,
and non-auditory physiological effects (EPA 1974). For
wildlife in natural environments, the most observable effects
of noise would seem to be changes in animal behavior due in
part to the masking of auditory signals. Most wildlife are
mobile and could in theory avoid areas of intense noise
levels necessary to cause permanent hearing damage.

It is well known that noise can cause great behavioral
changes in wildlife. Beginning with the human voice and
the clap of hands, and progressing to explosives and more
recently to sophisticated sonic generators, man has used
noise to repell animals from areas where their presence results in damage to crops or other property (Frings 1964,
Busnel and Giban 1968, Stewart 1974).

A report published in 1971 by Memphis State University
summarized much of what was then known of the effects of
noise on wildlife and other animals (Memphis State Univ.
1971). Of the 103 references cited in the report only one
made any reference to the effects of power line noise on
wildlife. This single report consisted of a brief mention
that the "hum" from power lines (no voltages or sound
levels given)adversely affected reindeer (Rangifer tarandus)
behavior and contributed to difficulties in herding (Klein
1971).

A report by the U.S. Environmental Protection Agency
(EPA 1974) identified a need for information on the effects
of noise on wildlife. Included in the report were recommendations for studies to determine the effects on
animals of low-level, chronic noise, and for comprehensive
studies on the effects of noise on animals in their natural
habitats. The subject of this paper is the possible effects

of relatively low-level transmission line AN on wildlife in natural environments.

This paper has the following objectives:

1) To describe the characteristics of transmission line AN and its effects on the human environment.
2) To relate these effects to the possible effects of AN on wildlife,
3) To relate the effects of AN to the overall effect of a transmission line and associated right-of-way, and
4) To describe research efforts of the Bonneville Power Administration (BPA) which are beginning to provide information for evaluating the effects of AN and other transmission line parameters on wildlife.

The effect of transmission line AN on wildlife is most reasonably considered when in the context of the overall effect that transmission lines have on wildlife. Noise from corona discharges is only one component of the complex environment that exists in the vicinity of a transmission line right-of-way. In addition to AN, wildlife on a right-of-way may be simultaneously exposed to human activity on access roads, shiny metal towers and conductors, electric and magnetic fields, and chemically-treated vegetation. It is a complex undertaking to determine the singular or combined impact of these components on various wildlife species which have widely differing biological and ecological characteristics which vary with season, weather, and habitat conditions.

TRANSMISSION LINE CHARACTERISTICS

An understanding of general transmission line characteristics is necessary to place the effects of AN in proper perspective. Although this discussion of transmission line construction, maintenance, and operation characterisitics relates primarily to the BPA transmission system, much of it will apply to transmission lines in general.

Transmission lines carry electrical power from generation sources to load centers. There the power is transformed to lower voltages in substations for distribution over smaller lines to users.

As an indication of the extent that transmission lines have become a part of our environment, consider that in 1970

there were an estimated 480,000 km of transmission lines in
the United States (USDI and USDA 1970). The BPA system
alone in the Pacific Northwest consists of over 20,000 km of
transmission lines (Figure 1). Of these over 5,600 km are
345 kV and above. It has been estimated that for the
balance of this century approximately 160,000 km of new
transmission lines will be constructed on 60,700 ha of right-
of-way each decade (USDI and USDA 1970). It is unlikely that
underground transmission will be used to any great extent in
the next few decades due to the high costs involved (Truax
1975).

Transmission structures vary greatly in shape and size
depending on such factors as line voltage, topography,
esthetics, and technical and safety design considerations
(BPA 1975). EHV transmission lines usually require metal
towers to support the weight of conducting wires (Figure 2).
In some cases one structure supports two electrical circuits. This results in a significant decrease in right-of-
way width required compared to that needed for two single
circuit lines but requires the use of taller towers.

The trend has been to build transmission lines with
higher operating voltages because the power carrying
capacity is significantly increased with increases in
voltages. For example, a 1200-kV transmission line could
carry approximately six times the power carried by a single
circuit 500-kV line but would require a right-of-way only
15.4 m wider (BPA 1976a). Presently, 765-kV is the highest
a-c voltage for operational transmission lines in the U.S.
BPA has constructed a 1100/1200-kV prototype transmission
line and lines of this voltage are expected to be in use in
the U.S. in the 1980's.

Construction

Transmission line construction usually involves vegetation clearing, access road construction, tower footing
installation, tower assembly and erection, conductor
stringing and site restoration (BPA 1974). The environmental
impact of this construction varies widely depending on such
factors as size and length of line, topography and vegetation
types encountered, weather, and time of year during which
construction occurs (Goodland 1973). In general, lines
constructed in steep forested areas where many new access
roads are required result in the greatest impact during the
construction phase. At the other extreme, lines constructed
through level grassland may result in comparatively few

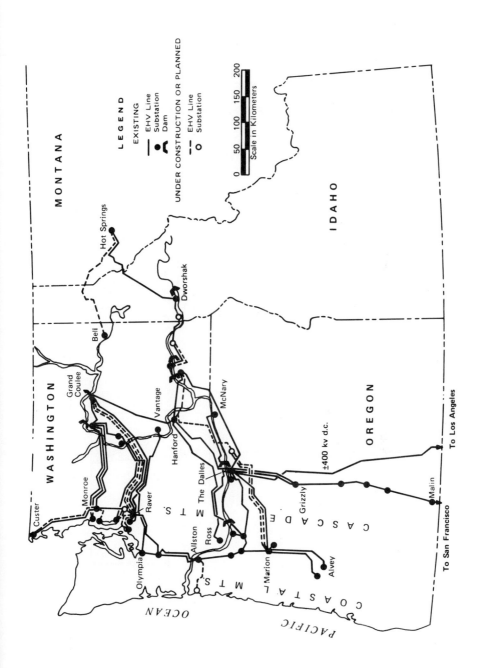

Figure 1 - Bonneville Power Administration Service area
showing E.H.V. transmission lines (345 kV, 500 kV, ±kv d.c.).

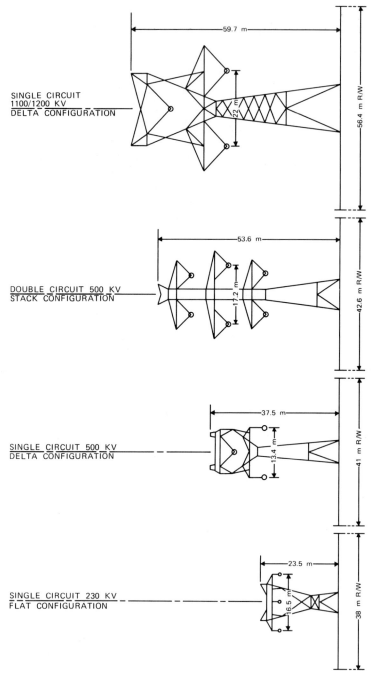

Figure 2 - Configurations of typical B.P.A. transmission structure. Scale is approximate.

environmental impacts during construction.

On older transmission lines a cleared right-of-way
through forest habitat is one of the most characteristic
features of a transmission line right-of-way. On some
newer lines more selective clearing is practiced resulting
in a "feathered" appearance.

Access roads can also be a primary feature of a transmission line right-of-way especially in forested areas.
Roads are required for both constructing and maintaining
transmission lines. Although existing roads are used where
possible new roads are often required both on and off the
right-of-way. Where access roads are constructed across
public lands such roads can receive heavy use by recrea-
tionists including hunters, campers, fishermen, and off-
road vehicle enthusiasts. Increased human access to previously remote areas can have significant consequences for
wildlife, especially those species that require wilderness
type habitat.

The final stage of construction consists of removing
all equipment, and material, and debris and restoring disturbed sites. Unmerchantable trees and brush removed from
the right-of-way and access roads are usually burned on
the construction site. Tower sites and other disturbed
areas are recontoured. Disturbed areas are often seeded to
facilitate the regrowth of vegetation.

As indicated above, construction of a transmission line
can result in significant environmental disturbances. The
resulting impact of construction on wildlife can be reduced
in some cases by the use of mitigation measures (BPA 1974,
USDI and USDA 1970).

Quantitative information on the above types of impacts
of construction activities on wildlife is rare. BPA line
construction personnel have reported that deer (Odocoileus
spp.) and elk (Cervus canadensis) are frequently observed
near construction sites. Other more secretive species may
avoid construction areas although during construction of the
BPA prototype 1100/1200-kV line a cougar (Felis concolor)
observed near the right-of-way did not appear to be greatly
disturbed by the activity of men and equipment. Construction
activity at the site did not appear to have any large effect on
on the abundance of birds and small animals. Stahlecker
(1975) reported that wildlife activity ceased when workers
were constructing a 230 kV transmission line in Colorado.
He also found that survey stakes, tower material and other

equipment were used as singing perches by some birds. He
reported that construction activity may have caused some
birds to desert their nests.   Other than possible adverse
effects on Ord's kangaroo rat (<u>Dipodomys</u> <u>ordi</u>) due to de-
struction of burrows, he found no evidence that other small
mammals were adversely affected by construction.   Balda
and Johnson (1976:7-18) reported that construction of two
500-kV transmission lines in Arizona resulted in drastic
differences in the species diversity of small mammals but
that the differences were transient in nature.

<u>Maintenance</u>

There occurs during the operating life of a transmission
line a variety of activities to maintain the facilities and
insure high system reliability. Of primary interest as far
as wildlife are concerned are vegetation management activities.   Brush and trees on rights-of-way are controlled
so that they will not grow into conductors or impede restoration of service if outages occur (BPA 1974).   Vegetation
management is accomplished in a variety of ways including
both mechanical cutting and chemical treatments.   The result, in forested areas, is that plant succesion is maintained in a shrub/grass stage.

Cleared rights-of-way are probably utilized by most of
the same wildlife species that inhabit the adjacent forest
(Cavanagh et al. 1976).   In some cases the right-of-way
may attract species and communities which differ from those
found in the original forest (Schreiber et al. 1976).   The
"edge effect" has been used to describe the phenomenon where
wildlife utilize areas of overlapping habitat types to a
greater degree than they utilize the adjoining habitats
(Odum 1959).   Several studies have documented the effects
that clearings, whether man made or natural can have on
wildlife (Lay 1938, Edgerton 1972, Walmo et al. 1972, Pengelly 1973, Patton 1974).

For transmission lines the width of the cleared right-
of-way can be an important consideration. Anderson et al.
(1977) found that in deciduous forests a 12 m wide right-
of-way had reduced bird species diversity compared to the
adjacent forest. A 30.5 m wide right-of-way had high bird
species density and diversity and a 91.5 m wide right-of-
way had less diversity but attracted species not found in
the adjacent forest.

Although herbicides are recognized as potentially

hazardous to wildlife, with proper application techniques they can be used effectively to control right-of-way vegetation while only minimally affecting non-target organisms (Buffington 1974). Since 1970 BPA has sponsored research to evaluate herbicides used on transmission line rights-of-way (Norris 1971). This research also shows that these chemicals can be used without noticeable adverse effects to animals. Buffington's (1974) review pointed out some possible effects of herbicides on wildlife which have not received a great deal of attention. These include the possible adverse effects of herbicides on rumen flora of large herbivores.

Mayer (1976) studied wildlife usage of power line (no voltages given) rights-of-way in the eastern U.S. which had been maintained by the use of herbicides. Wildlife species included whitetail deer(Odocoileus virginanus) ruffed grouse (Bonasa umbellus) gray squirrel (Sciurus carolinensis) bobwhite quail (Colinus virginianus), and cottontail rabbit (Sylvilagus floridanus). Results of the study showed that wildlife utilized the rights-of-way to a greater degree than they utilized habitats adjacent to the right-of-way.

Vegetation on rights-of-way can also be controlled by selective cutting and this can also have an effect on wildlife usage. Cavanagh et al. (1976) found that wildlife usage of selectively maintained powerline (no voltage given) rights-of-way was higher than usage of clearcut rights-of-way. Costs for selectively maintaining the rights-of-way was approximately 12 times that required for maintaining the clearcut rights-of-way.

Operation

The electrical properties of an operating transmission line result in field effects and corona effects. A consideration of the electrical effects of transmission lines is necessary in both the design of biological studies and in the interpretation of results. The following is an introduction to the electrical characteristics of EHV transmission lines. Persons unfamiliar with transmission lines contemplating biological research in this area should seek advice and assistance from persons with training in such fields as electrical engineering and physics.

Field Effects

Voltage applied to a transmission line conductor produces an electric field in the region surrounding the con-

ductor and extending to the earth. Current flowing in a
conductor produces a magnetic field. At a common measurement height of 1 m above ground, the maximum electric field
strength (voltage gradient) beneath BPA 500-kV lines with
a conductor to ground clearance of 11 m is about 8-kV/m
(BSTT 1977). The electric field strength near an a-c transmission line can be measured with a hand-held meter available
from various manufacturers. Field strength can also be computed (Deno and Zaffanella 1975). The maximum magnetic
field strength of such lines is approximately .6 Gauss
(BSTT 1977). The electric field strength beneath 765-kV
lines, with a conductor to ground clearance of 17 m is 9-10
kV/m (SNYPC 1976). These maximum field strength levels
occur in a relatively small area at mid span beneath the
conductors. At the edge of a 500-kV transmission line right-
of-way maximum field strength is from 2.5 - 3.5 kV/m (BSTT
1977). Maximum field strength beneath BPA 230 kV lines is
3-4 kV/m. For reference, the electric field strength 30
cm from an electric blanket is approximately 0.25 kV/m and
60 Hz electric fields inside a typical house may range from
less than 0.001 kV/m up to 0.013 kV/m (Miller 1974).

When conducting objects such as vehicles, persons, or
animals are in an electric field, currents and voltages are
induced in them (BSTT 1977). Usually these currents and
voltages are below the perception level for humans. Under
certain circumstances annoying spark discharge shocks can
occur to people and animals in the vicinity of transmission
lines. These circumstances occur when a person or animal,
insulated from ground, comes in contact with a grounded
object, or when the object is insulated and the person or
animal is grounded. Such shocks are similar to what one
experiences after walking across a carpet and touching a
door knob. Conducting objects near transmission lines such
as metal fences are routinely grounded to prevent the build-
up of large voltages on such objects (BSTT 1977). The
magnetic field can also induce voltage and currents although the effects are not as apparent as for the electric
field.

The presence of electric fields can also be sensed if
the magnitudes are great enough. Deno and Zaffanella (1975)
reported the threshold for perception of an electric field
(e.g., hair stimulation) for the most sensitive 10 percent
of the persons tested was between 10 and 15-kV/m. We have
felt hair stimulation on our extended arms beneath a BPA
500-kV line with an estimated field strength of 7-8 kV/m.
When standing on dry ground with rubber soled shoes, we
could also perceive a slight tingling sensation when touching

vegetation.

Relatively little is known about the perception of animals to electric fields. In one study there were no visible changes in grazing, feeding, and drinking habits of cattle on damp ground in electric fields of 18-kV/m (Deno and Zaffanella 1975). Those authors, however, suggested that on dry ground, spark discharges might occur between grounded objects and the bodies of animals. Researchers at Battelle Pacific Northwest Laboratories saw no hair movement on the ear of an anesthetized swine until field strength reached 50-55 kV/m (Phillips et al. 1976).

Radar tracking studies conducted by Larkwin and Sutherland (1977) suggested that during nocturnal migratory flight, birds were apparently able to sense low intensity a-c (72-80) electromagnetic fields produced by a large antenna system. The two antennas used in the study were each 22.6 km long and 8 m above ground. They produced an electric field of 0.07 V/m at 100-400 m and the magnetic field at these distances was less than 1 percent of the earth's magnetic field.

Lott and McCain (1973) used implanted electrodes to determine if rats were aware of an external electric field. When in a d-c positive field of at least 10 kV/m rats showed a statistically significant increase in brain activity (EEG) which was reported as an indication the rats were aware of the field.

In recent years considerable interest has arisen over the question of whether low intensity electric and magnetic fields can result in biologic effects as a result of long term exposure (Young and Young 1974, Llaurado et al. 1974). At the present time the bulk of the available evidence does not indicate that transmission lines pose a significant biological hazard in this regard (Bridges 1975, BSTT 1977, Janes 1976). Additional information is needed, however, for assessing the potential for such effects.

In addition, studies done on the biologic effects of ions at concentration levels similar to those produced by a d-c transmission line suggest both beneficial and adverse effects are possible (Krueger and Reed 1976). Several studies are underway in the U.S. which should provide more definitive information on the nature and significance of biologic effects from electric and magnetic fields. A review of the above topics and listing of this research can be found in a BPA publication (BSTT 1977).

A d-c transmission line has similar electrical parameters as an a-c line, however, there are some differences.
Corona-generated ions from a d-c transmission line form a
"space charge" which alters the electric field near the line
(Hill et al. 1977). Ions are greatly affected by wind and
consequently the ground level electric field for a d-c line
frequently changes in location and magnitude. On the Celilo-
Sylmar $\pm$ 400-kV d-c line electric field strengths of -34 kV/
m have been measured with approximately half of this due to
enhancement by the space charge (Bracken et al. 1977).

For reference, the earth's d-c electric field averages
0.13 kV/m and beneath thunderclouds levels of 3 kV/m or
higher may exist even in the absence of lightning (Polk
1974). The magnetic field of a d-c line is approximately
the same level as the .6 G d-c field of the earth.

It should be pointed out that d-c electric field
strength values cannot be directly compared with a.c. values.
Even with d-c electric fields of 40 kV/m the current inter-
cepted by persons beneath a d-c tramsmission line is many
times below the perception level (Hill et al. 1977). The
probability of receiving perceivable spark discharge shocks
in the proximity of a d-c line is also less than for comparable a-c lines (Hill et al. 1977).

Audible Noise (AN)

When the electric field intensity on the surface of
a transmission line conductor exceeds the breakdown strength
of air, corona dischrages occur (Deno and Comber 1975).
Corona results in AN, radio and television interference,
flashes of light, and production of oxidants (primarily
ozone). Of these, AN is probably the more important as
far as possible effects on wildlife are concerned. Studies
have shown that the amount of ozone produced by transmission lines is generally not measurable above ambient
levels (Sebo et al. 1976, Roach et al. 1977).

The subject of transmission line AN has been discussed in a number of papers (Perry 1972, Ianna et al. 1973,
Comber and Zaffanella 1975). What follows is a summary
of pertinent aspects of corona produced noise which provides a basis for determining the possible effects of such
noise on wildlife.

Depending on such factors as line voltage, conductor
design and surface irregularities, and weather, transmission

line AN can vary widely. It is most noticeable on a-c lines of 500 kV or higher. For such lines the noise is characterized by a random broadband, crackling, hissing sound with a 120 Hz "hum" or harmonics of this frequency occasionally present. The characteristics of AN from d-c transmission lines is described later in this section. Although usually termed "audible noise", corona noise actually extends to frequencies above the limits of the human hearing range (Ianna et al. 1973).

Corona discharges result in power losses and therefore a-c transmission lines are designed to operate in fair weather below the corona onset level. A common design is the use of a number of subconductors to transmit the power in each of the three phases of an a-c transmission line or the two poles in the case of a d-c line. This has the effect of increasing the overall surface area of the conducting material. This reduces the voltage gradient on the conductor surface and thus reduces the occurrence of corona.

In reality corona can occur with EHV lines even during fair weather because of imperfections or contaminants on the conductors. Although nicks, scrapes, insects, and dust can cause corona, water on the conductors is the most important cause. A-c transmission line AN is generally highest during heavy rain. Ambient noise, however, is also high during heavy rain so AN is more apparent during snow, fog, light rain, mist, or just after a rain while the conductors are still wet.

The age of the conductors also has an effect on the amount of AN produced. The surface of a new conductor has a light coating of grease which causes water droplets to form over the entire conductor surface (Perry 1972). As the conductor ages, corona and weather combine to change the conductor surface conditions so that water droplets tend to form only on the underside of the conductor. This reduces the effective number of corona discharge points and lowers AN compared to the new conductor. AN may reach its highest levels during the first few years after a line is constructed which is also the time when the "newness" of the overall facility may have its greatest effect on wildlife.

Other factors which can affect AN production include the configuration of the subconductor bundles, phase spacing and phase configuration, height of the conductors above the ground, and proximity of other circuits which may be on the same transmission structure or on adjacent ones.

Analysis of transmission line AN should include both the broadband and the pure tone components. The 120 Hz hum (and other multiples of this tone) is produced by the rapid oscillation of ionized air ions near the conductor surface which have been generated by corona. Although the 120 Hz hum correlates with corona, the relative magnitude of the random noise and the 120 Hz hum can differ depending on weather conditions. The 120 Hz hum does not attenuate as rapidly as the high frequency random noise and it therefore may be detected at greater distances from the transmission line. At certain times in central Oregon with ambient noise levels of approximately 20 dB(A) we have heard the hum from two 500-kV a-c transmission lines from almost 2 km away.

A standard procedure for making measurements of transmission line AN has been developed (IEEE Committee Report 1972). Measurements of AN are often reported in dB(A). The "A" weighting network discriminates against lower frequencies including 120 Hz, so dB(A) measurements of AN actually pertain primarily to the random noise (Comber and Zaffanella 1975).

Figure 3 shows a lateral profile of AN measurements for a 500-kV line during rain.

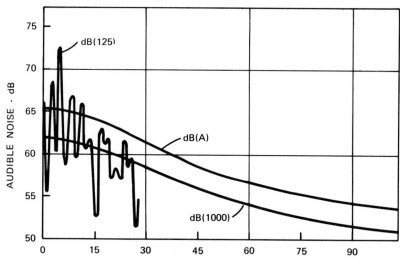

Figure 3 - Typical lateral profile of audible noise from a 500 kV a-c transmission line with one 6.35 cm conductor for each phase. Adapted from Perry (1972).

Note that AN levels measured at various distances from a transmission line are functions of distance and the combination of random noise and hum generated from each phase of the line. The strength of the 120 Hz hum can vary several dB as a consequence of displacing the measuring microphone a few meters.

In determining the environmental effect of transmission line AN it is important to consider it in relation to ambient noise levels. If the AN measured for a transmission line is at least 10 dB above ambient, the measured AN is essentially from the line (Comber and Zaffanella 1975:197). Those authors further indicate that when differences measured between ambient and the transmission line are 3 dB or less the two conditions cannot be separated by conventional measures. It should be pointed out that the above refers to measurable levels. The AN may still be audible even when the level is similar to ambient due to its unique frequency characteristics.

As described above, AN varies greatly in intensity and frequency composition depending on a number of factors foremost of which is weather. In considering the possible effects of AN on wildlife, it is important to keep in mind that animals that may inhabit areas near an a-c transmission line right-of-way are not exposed to a constant level of noise. Figures 4 and 5 are examples of the kinds of variations in AN which can be expected from EHV transmission lines.

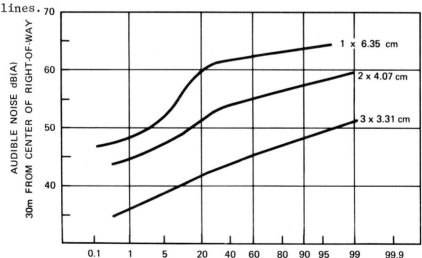

CUMULATIVE PERCENT OF TIME LESS THAN ORDINATE

Figure 4 - Cumulative frequency distribution for audible noise from a 500 kV a-c transmission line with one 6.35 cm conductor for each phase. Adapted from Perry (1972).

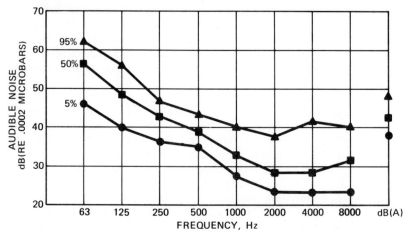

Figure 5a - Audible noise spectrum during fair weather or a
775 kV a-c test transmission line with four 35.1
mm diameter conductors per phase. Noise level
is less than the ordinate the indicated percentage of time. Redrawn from Kolcio et al. (1973).

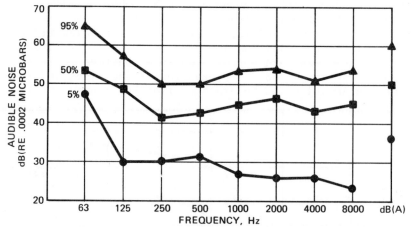

Figure 5b - Audible noise spectrum during fog for a 775 kV
a-c test transmission line with four 35.1 mm
diameter conductors per phase. Noise level is
less than the ordinate for the indicated percentage of time. Redrawn from Kolcio et al. (1973).

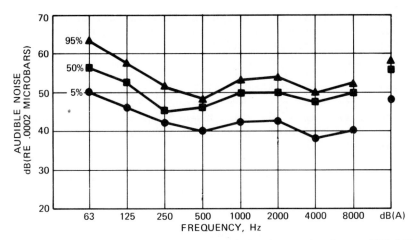

Figure 5c - Audible noise spectrum during rain for a 775 kV
a-c test transmission line with four 35.1 mm
diameter conductors per phase. Noise level is
less than the ordinate for the indicated percentage of time. Redrawn from Kolcio et al. (1973).

Although there is no difference in the mechanism of a-c
and d-c corona, there are sufficient differences in AN
associated with the two types of transmission lines to war-
rant a brief discussion. This discussion is based primarily
on information contained in a recently published reference
book on HVDC transmission (Hill et al. 1977).

Unlike three phase a-c lines, a d-c transmission line
has only two sets of conductors (poles), one negative and
one positive. The positive pole is the primary source of
AN. The noise is characterized by impulsive pops and is
similar to the random noise of the a-c line without the
120 Hz hum. The frequency characteristics of the d-c AN
are generally similar to the a-c random AN.

A major differnce between the two types of lines is that
d-c corona loss is less affected by weather than is a-c.
Correspondingly, as compared to a-c lines, AN from a d-c line
changes less between fair and inclement weather conditions.
Studies with a d-c test line at the Dalles, Oregon showed
that rain may even cause a very slight decrease in d-c AN
levels and snow had no significant effect (Hill et al. 1977:
67). Maximum AN levels measured for the d-c test line
ranged from 30 dB(A) with the line operating at $\pm$ 400 kV,
to 48 dB(A) at $\pm$ 600 kV. Note that for a d-c line, voltage
is usually reported as the difference in potential between
the positive and negative poles and ground. Sometimes the

operating voltage is reported as the total potential between the two poles.   In this latter case $\pm$ 400 kV d-c would be reported as 800 kV d-c.

To gain a better understanding of how AN from transmission lines could affect wildlife one can first consider its effects on people.   For reference, Table 1 shows levels of familiar noises and gives human responses to such noises. AN can be a source of annoyance to persons living near EHV transmission lines.

The concept of "annoyance" in regards to wildlife is not well defined.   Stewart (1974) whose company manufactures sonic animal repellent devices believed that humans and lower animals have much in common regarding their responses to noise.   Consequently the widely use "Av-Alarm" sonic generator successfully used in bird and mammal control work was designed to produce the kinds of noise that promote annoyance, nervousness, and discomfort in people (Stewart 1974).

One measure of the degree of human annoyance to transmission line AN is the number of complaints a power company receives.   A guideline to the probability of receiving complaints about AN was developed by Perry (1972).   When AN levels 30 m from the center of the right-of-way are approximately 53 dB(A) or lower no complainst would be expected.   AN above 59 dB(A) can result in numerous complaints. Perry (1972) cautioned that in practice the "acceptability" of AN of various levels varies depending on such factors as ambient noise, population density and level of 120 Hz or other tones present in the AN.

Because of problems with audible noise BPA no longer builds new 500-kV lines with a single subconductor per phase and some lines of this type have been reconductored to the newer three subconductor per phase design.

The subject of AN was one of the central issues in extensive hearings held by the State of New York Public Service Commission for the purpose of certifying the construction of 765-kV transmission lines in that state. The hearings began in 1974 and by June 1976 the testimony of over 26 scientists and engineers had filled more than 10,000 pages (SYNPSC 1976).   In addition to AN, testimony was taken on ozone, induced electric shocks, and on the biological effects of electric fields.

Table 1. Examples of Environmental Noise Levels and Typical
Human Responses. (Source: EPA 1972)

	Noise Level (Decibels)	Human Responses	Conversational Relationships
	150		
Carrier Deck Jet Operation	140		
		Painfully Loud	
	130	Limit Amplified Speech	
Jet Takeoff at 61 m	120		
Discotheque		Maximum Vocal Effort	
Auto Horn at .9 m			
Riveting Machine	110		
Jet Takeoff at 610 m			Shouting in ear
Garbage Truck	100		
N.Y. Subway Station		Very Annoying	
Heavy Truck at 15 m	90	Hearing Damage (8 hours)	Shouting at .6 m
Pneumatic Drill at 15 m Alarm Clock	80	Annoying	Very Loud Conversation at .6 m
Freight Train at 15 m			
Freeway Traffic at 15 m	70 1/	Telephone Use Difficult	Loud Conversation at .6 m
Air Conditioning Unit at 6 m	60	Intrusive	Loud Conversation at 1.2 m
Light Auto Traffic at 30 m	50	Quiet	Normal Conversation at 3.7 m
Library	40		
Soft Whisper at 4.6 m	30	Very Quiet	
Broadcasting Studio	20		
	10	Just Audible	
	0	Threshold of Hearing	

1/ Contribution to hearing impairment begins.

The proposed 765-kV transmission lines considered in the hearings would produce a maximum AN level less than 60 dB(A) at the edge of a 76 m wide right-of-way. Sleep interference was the most serious possible effect predicted for AN. Although the testimony in the hearings dealt primarily with humans, at least two witnesses addressed possible effects on wildlife (Driscoll 1976, Fletcher 1976). Fletcher (1976) suggested that AN from the proposed 765-kV lines could at times mask certain acoustic signals that could be critical to the survival of some animals. He added that such effects on wildlife, however, would probably be insignificant. Further, scientific evidence on which to base those predictions were almost completely lacking according to Fletcher. Driscoll (1975) believed AN could affect wildlife breeding activities and predator-prey relationships. He suggested the right-of-way may be less desirable to wildlife during inclement weather because of high levels of AN. He also felt that little information was available to permit quantitative assessment of the effects of AN on wildlife and that research on the subject was needed.

One of the major difficulties in determining human annoyance levels associated with AN is that annoyance is determined by a number of psychological factors. Anticaglia (1970:2) pointed out that evaluation of noise by people involves both the consideration of the physical sound and the subjective impression in the listener's mind. For example any amount of AN may be extremely annoying to a person when it is coming from a transmission line which was constructed across his property against his wishes and only after a lengthy condemnation proceeding.

Most of the literature dealing with the biologic effects of transmission line AN is oriented toward possible effects on humans. In relating this literature to wildlife some distinctions should be made. One of the main distinctions is that AN as usually reported for transmission lines is for locations at the edge of the right-of-way or beyond. Easements purchased for rights-of-way are for the purposes of operating and maintaining the transmission line so persons who might find the higher AN levels on the right-of-way objectionable are not obliged to stay there. Wildlife of course are not aware of the existence of legal right-of-way boundaries. In considering the possible effects of AN on wildlife it is therefore necessary to know the maximum AN levels which occur on the right-of-way as well as levels at various lateral distances. As previously mentioned AN levels can change significantly within short distance from the line.

A related factor is that people are usually inside closed buildings during inclement weather when a-c transmission line AN is highest. Wildlife, however, may still be moving about and have greater opportunity to encounter the high levels of AN. Unfortunately, relatively little has been reported in the literature about the effects of weather on various wildlife species at different seasons of the year. Species that become dormant during winter or that migrate may encounter high levels of AN less frequently than resident species. Rain, wind, and snow also adversely affect an observer's ability to detect and study animals (Overton and Davis 1969:425).

Another distinction is that AN measured in dB(A) may not be the most appropriate measure of the noise as it is perceived by wildlife. Although some species, including most birds, have a hearing range which approximates that of man other species may be sensitive to a much wider range of frequencies. For a-c lines at least it is important to consider the tonal component of the noise including frequencies outside of those normally considered of importance to man.

Table 2 shows frequency hearing ranges of various animals as compared with man. Many insects respond to frequencies far above those audible to humans. The hearing of most birds is similar to man and although many bird vocalizations contain ultrasonic frequencies birds probably cannot hear such frequencies (Sales and Pye 1974). Mammals respond to a wide range of frequencies including those audible to man and those in the ultrasonic range. In reptiles the importance of sound perception is subordinate to vision and chemoreception and sound producing mechanisms are absent in most reptiles (Bogert 1960).

As mentioned previously corona can produce ultrasonic frequencies. We are not aware of any study in which ultrasonic frequencies were measured in the corona noise produced by a transmission line under normal operating conditions. Ianna et al. (1973) studied the spectral characteristics of corona produced by metallic protrusion on conductors in a laboratory. In one test with a voltage gradient on the conductor of 13.3 kV/cm negative dc, the peak of the acoustic spectra was between 24-26 kHz.

Although it is reasonable to assume that transmission line corona could produce ultrasonic frequencies which could be heard by certain wildlife species, it is less likely that such frequencies have any significant effect except very near the conductors. Ultrasonic frequencies are rapidly at-

Table 2. Hearing Abilities (Frequencies) of Various Animals as
Compared With Man.

Species	Lower Limit (Hz)	Maximum Sensitivity (Hz)	Upper Limit (Hz)	Source
Man (Homo sapiens)	16	4,000	20,000	EPA (1974)
		Invertebrates		
Tiger moths [1] (Arctiidae)	3,000	--	20,000	Haskell and Belton (1956)
Noctuid moth [1] (Prodenia evidania)	3,000	15,000-60,000	240,000	Roeder and Treat (1957)
Butterflies (38 species) [1] (Lepidoptera)	--	40,000-80,000	--	Schaller and Timm (1949,1950) cited in Autrum (1963)
Long-horned grasshoppers [1] (Tettigoniidae)	800-1,000	10,000-60,000	90,000	Wever and Vernon (1957)
Field cricket [1] (Gryllus)	300	--	8,000	Wever and Bray (1933)
Mosquitoe [2] (Anopheles subpictus)	150	380	550	Tischner (1953), cited in Autrum (1963)
Male Midges [2] (Tendipedidae)		80-800 with peaks at 125 and 250		Frings and Frings (1959)
		Amphibians		
Bullfrog (Rana catesbeiana)	<10	<1,800	3,000-4,000	Strother (1959)
		Birds		
Starling (Sturnus vulgaris)	<100	2,000	15,000	Granit (1941), cited in Bremond (1963)
House sparrow (Passer domesticus)	--	--	18,000	Granit (1941), cited in Bremond (1963)
Crow (Corvus brachyrhynchos)	<300	1,000-2,000	>8,000	Trainer (1946)
Kestrel (Sparrow Hawk) (Falco sparverius)	300	2,000	>10,000	Trainer (1946)

Table 2. Hearing Abilities (Frequencies) of Various Animals as
          Compared With Man. (cont.)

Species	Lower Limit (Hz)	Maximum Sensitivity (Hz)	Upper Limit (Hz)	Source
Long eared owl (Asio otus)	<100	6,000	18,000	Schwartzkopff (1955)
Mallard duck (Anas platyrhynchos)	300	2,000-3,000	>8,000	Trainer (1946)
Mammals				
Bats (Chiroptera)	<1,000	30,000-100,000	150,000	Sales and Pye (1974)
Rodents (Rodentia)	<1,000	5,000-18,000, and 40,000-60,000	100,000	Sales and Pye (1974)
Cats (Felidae)	--	--	70,000	Evans (1968) cited in Sales and Pye (1974)
Opossum (Didelphus virginiana)	<500	--	>60,000	Sales and Pye (1974)

1/ Frequencies of continuous tones that stimulate the tympanal organs.
2/ Frequency response of Johnston's Organ which are located at the base
    of the antennae.

tenuated especially by fog (Sales and Pye 1974). Those authors also suggested that bats seem to avoid flying during fog because these animals utilize ultrasonic echo location for navigation. Atmospheric moisture, while causing production of corona and possibly ultrasonic frequencies, also acts to decrease the propagation of these sounds.

Another point pertains to the ambient noise levels with which transmission line AN levels are compared. Even when considering the possible effects of AN on people the significance of extremely low ambient levels as a basis for comparison of the AN is sometimes overlooked. The range of environmental sound levels in the U.S. is very great. The range of day-night (L_{dn}) sound levels extend from the region of 20-30 dB for wilderness areas to the region of 80-90 dB or higher in noisy urban areas (EPA 1974). It should be pointed out that even sounds in the natural environment can reach high levels. Griffin (1976) measured sound levels near a frog pond on a still night and found that steady frog calling produced noise levels of 55-60 dB(A) with peaks to 75 dB(A). Waterfalls may produce noise levels of 85 dB(A) or more (EPA 1974).

In contrast, on a number of occasions we have measured sound levels as low as 15-20 dB(A) in central Oregon in places far from human development. At such low levels one strains to hear the slightest sound. Even light to moderate levels of transmission line AN contrasted to these conditions can be annoying.

Frings (1964) reported that in pest control applications the use of amplified communication signals as low as 3 dB above ambient can cause reactions in birds. When male Japanese quail (<u>Coturnix japonica</u>) were exposed for two hours to white noise at levels of 12 dB(A) above ambient the birds significantly increased the frequency of their separation crowing (Potash 1972). Such a response to increases in ambient noise may aid a separated pair of quail in re-establishing contact. As Potash (1972) pointed out, however, the increased chance of being detected by the prospective receiver must be weighed against the chance of being detected by a predator. In another case, Frings and Frings (1959) found that small male flies (<u>Pentaneura aspera</u>) responded with an agitated circling and gathered around the sound source when it was producing 125 Hz tones 13-18 dB above ambient or 250 Hz at 3-8 dB above ambient.

The fact that noise from electrical devices can attract insects has been noted by others (Sotavalta 1963:387).
This raises the possibility that although transmission line
AN may repel some species, it may attract others. Also,
insects attracted to the right-of-way could also influence
the density and distribution of insectivorous birds in the
vicinity of transmission lines.

Another point regarding distinctions between human and
animal responses to transmission line AN pertains to the
meaning of noise. As previously mentioned, noise is usually
defined as something unwanted. From the human viewpoint it
is difficult to imagine any positive effects of transmission
line AN. For wildlife, sounds from the inanimate environment
are potential carriers of useful information (Emlen 1960:xi).
For some kinds of wildlife the possibility exists that the
sound produced by a transmission line could act as a navigational aid. Although this suggestion is highly speculative
at this time we feel it is worthy of consideration in light
of recent research. In one report, Griffin (1975) presented
evidence which suggested that on cloudy or foggy nights
migrating birds may obtain navigational information from
sounds characteristic of particular environments. Yodlowski
et al. (1977) reported that homing pigeons can detect sound
energy below 10 Hz at amplitudes within the range of those
occuring in the environment. "Properly utilized, infrasound information could thus assist in almost every aspect
of avian navigation, in both homing and migration" (Yodlowski
et al. 1977:226). Birds have been known to utilize transmission line corridors as travel lanes (Carothers and Johnson
1975:215, Grue 1977:214). During inclement weather could
sounds produced by corona along a north-south running transmission line similarly aid nocturnal avian migrants?

Before concluding our discussion of AN we would like to
mention one other topic which is important when considering
the possible effect of AN on wildlife. Throughout this paper
we stress that AN is only one component of a transmission
line environment. The effects of AN then must be viewed in
relation to the overall effects of the line. The combined
effect of AN with effects from other transmission line parameters could take several forms. Possibilities for combined effects in such situations could be indifferent, ad-
ditive, synergistic, or ameliorative (EPA 1974:E-3). The
effect of relatively low level AN if it combined synergistically with the effect of another parameter could result in
a greater effect than if AN were considered alone. Certain
chemical agents, and certain vibrations can have synger-
gistic effects when combined with noise (EPA 1974:E-5). In

bird control work (Busnel and Giban 1968) felt the addition
of a visual stimulus would enable lower sound intensities
to be used when using sonic repellents. In the transmission
line environment the effects of AN may combine with the
electric and magnetic field synergistically.

THE EFFECTS OF TRANSMISSION LINES ON WILDLIFE

Although EHV transmission lines were first put into
operation in the 1950's, it was several years before any work
was done to systematically document the effect of these lines
on wildlife. Passage of the "Environmental Policy Act of
1969" with its requirements that Federal agencies prepare
environmental impact statements for all major projects
pointed to the need for additional data for predicting environmental impacts.

One of the first attempts to assess the state of knowledge of the environmental impact of transmission lines was
a colloquium, "Biotic Management Along Power Transmission
Rights-of-Way" held at the University of Massachusetts in
June 1973 (Goodland 1973). Another contribution which
further defined the environmental impact of transmission
lines was published a year later by the Oak Ridge National
Laboratory (Kitchings et al., 1974). The most recent
national event to provide a forum for discussing the impact
of transmission lines on wildlife was a symposium on "Enviro-
mental Concerns in Rights-of-Way Management", held in
January 1976 at Mississippi State University (Tillman 1976).

Biological Research on the BPA Transmission System

Having briefly described the transmission line enviro-
ment and potential sources of impact on wildlife we will now
describe BPA research conducted to date. This research has
provided information with which to begin assessing the possible effects of transmission line AN, and other parameters,
on wildlife.

BPA is a Federal agency responsible for marketing the
power generated by all Federal dams in the Columbia River
Basin. BPA transmission lines are located in a wide variety
of wildlife habitats including some of the most productive
big game habitat in the U.S. In 1974 BPA developed a program to begin determining the effects of transmission lines
on wildlife.

Idaho 500-kV a-c Transmission Line Study

The first research project developed in the BPA program
was designed to determine the effects of a 500-kV transmission line on migrating Rocky Mountain elk (Cervus canadensis nelsoni) (Lee 1974). Concerns have been expressed that
transmission lines could interfere with movement of migratory
wildlife species (Villmo 1972, Martinka 1974, BPA 1974).
Villmo (1972:8) reported that the unusual noise generated by
power lines (no voltages or noise levels stated) seemed to
frighten reindeer in Scandinavia. He stated that "When
herders are attempting to move a reindeer herd across a power
line which is generating noise we know that the reindeer react to the sound and are reluctant to pass under the lines"
(Villmo 1972:8). He also said that in some cases it could
not be determined whether the power line itself or the
cleared right-of-way was alarming the reindeer.

BPA sponsored an intern with the Western Interstate
Commission for Higher Education (WICHE) to conduct a 1-year
study of the possible effects of a transmissiuon line on
elk migration. The study was completed in June 1975 and was
one of the first studies to systematically investigate and
document wildlife behavior near a 500-kV line (Goodwin 1975,
Goodwin and Lee 1975).

The study was designed to test the hypothesis that elk
movement on, and use of a 500-kV transmission right-of-way,
is no different than that occurring on other forest
clearings. The study area was in northern Idaho along the
BPA 500-kV line between Dworshak Dam and Hot Springs, Montana (Figure 1). The line is located in heavily forested
areas of the western red cedar (Thuja plicata)/ western
hemlock (Tsuga heterophylla) vegetation type.

Construction of the Hot Springs-Dworshak line began
in July 1968 and was completed in October 1972. The line
was engerized in March 1973. Voltage extremes on the line
range from 500-550 kV. The steel, delta configuration towers
average 37 m in height and are spaced approximately three per
km. Three subconductor configurations are used on the line;
the single, double and triple subconductor per phase forms.
Audible noise characteristics for these configurations vary
considerably (Figure 4).

The study was based on a comparison of animal movement on and use of the right-of-way with that on other
clearings. Five paired right-of-way and control areas were
used during the study. Study methods included direct ob-

servations, time lapse photography, track counts, and vegetation analyses. In addition to elk, the study provided information on several other wildlife species, and on the use
of the right-of-way by hunters.

When an animal was observed in the right-of-way or control, an audible noise measurement in dB(A) was made under
the outer conductor and the noise level at the animal's location was estimated. Estimates were made based on the knowledge that line noise attenuates at a rate of 3 to 4 dB for
each doubling of distance (Perry 1972). Noise measurements
were made with a General Radio Type 1551-C sound level meter.
The 25 mm ceramic microphone was approximately .5 m above
the ground when measurements were made. No measurements of
the 120 Hz hum or other frequencies were made.

Between August and December 1974 a total of 310 hours
was spent observing two paired right-of-way and control
areas. Due to minimal use which the study areas received
during this time of year and the nocturnal behavior of many
animals, very few observations were made. Only 4 deer, 2
bear (Ursus americanus), 6 elk, and 22 coyotes (Canis latrans) were observed. From March to May 1975 less emphasis
was placed on direct observations, however, several hundred
deer and elk were seen while the observer was counting tracks
and servicing time lapse cameras. The problems which limited
the direct observations in the fall also applied to the
time lapse cameras.

Typical behavior of most deer and elk when they entered
the right-of-way or control was to remain motionless for a
few moments at the forest edge. After this, they entered the
clearing and usually began feeding. After initial wariness
when entering a clearing none of the deer or elk observed
appeared to be disturbed by the presence of the transmission
line.

Results of the track counts indicated deer and elk
movement across all forest openings during the fall was low,
but increased steadily throughout the season. Elk movement particularly increased once snow began to accumulate.
The tracks of 87 elk and 9 deer in the vicinity of the right-
of-way were followed in the snow. Thirty-eight percent of
the animals maintained a relatively straight line of travel
as they approached, crossed, and left the right-of-way.
Sixty percent of the animals followed roads or established
game trails in crossing the right-of-way. Only 2 percent
failed to cross the right-of-way. In these two cases elk
came within 25 m of the right-of-way and appeared to have

come to the area to feed.

Generally the right-of-way and control areas had significantly greater ground cover than the adjoining areas. The percentage of grasses, forbs, and shrubs sometimes differed considerably between right-of-way and control.

Fair weather AN on the right-of-way for the single conductor per phase configuration ranged from 45 to 55 dB(A). Levels of 62-68 dB(A) were measured during rain. For the two conductor per phase configuration, maximum fair weather AN was 35 to 40 dB(A). AN from the three conductor per phase configuration was usually less than 30 dB(A). A reading of 52 dB(A) was recorded on the right-of-way during a light shower. AN measured in the control areas was usually below 30 dB(A). The highest reading in a control area was 44 dB(A) which was during a light rain.

During several days of 55 to 60 dB(A) noise levels, deer and elk track counts did not indicate any aversion by the animals to the right-of-way. The study was not specifically designed to provide quantitative data on the possible effects of the line on birds, however, ravens, grouse, and several other bird species were observed near the line when noise levels exceeded 60 dB(A). During the study a herd of approximately 60 bighorn sheep was observed in the Dworshak-Hot Springs 500-kV right-of-way near Hot Springs, Montana. During a light rain with AN of 53 dB(A) the sheep were observed bedded down on the right-of-way.

The activity of hunters on the right-of-way and access roads had a significant influence on elk movement. Although segments of the right-of-way were closed by locked gates, some hunters broke the gates in order to drive onto the right-of-way. Many persons hunted the right-of-way on foot and some used motorcycles. Hunting pressures had a definite effect on elk distribution. Elk moved away from hunting pressure and concentrated in places 0.8 km or more away from roads and clearings.

Data obtained during the Idaho study indicated that elk movement near the 500-kV transmission line right-of-way was not significantly different than that near other forest openings. The transmission line produced audible noise levels which at times were very annoying to humans. There was no indication, however, that the noise deterred elk, deer and several other wildlife species from entering and crossing the right-of-way. Elk and deer use of the right-of-way and control areas was primarily a function of available forage.

Elk and deer avoided the right-of-way and control clearings during the hunting season. The transmission line, however, did not appear to prevent these animals from eventually crossing the right-of-way and continuing their migration.

1100/1200-kV Transmission Study

In April 1976 another BPA-sponsored biological study began. This one is being conducted at the site of the BPA 1100/1200-kV prototype transmission line near Lyons, Oregon (Figure 6). This 31-month long study is being conducted by Battelle-Northwest of Richland, Washington.

During the Environmental Impact Statement process for the 1100/1200-kV project, questions were raised as to the possible biologic effects of such a line (BPA 1976b). No significant adverse biologic effects are expected. However, because of the concerns which were raised, and because few scientific studies have been conducted for existing trans- mission lines, BPA decided biological studies would be an integral part of the prototype project.

Construction of the 2.1 km long line began in the spring of 1976 and was completed the following fall. The line was constructed on a right-of-way which had contained three 230-kV transmission lines. One of the 230-kV lines was removed and the 1100/1200-kV was constructed in its place and relatively little tree clearing was required. The first of three transformers was energized December 21, 1976. All three phases were energized on May 10, 1977.

The prototype line is providing a source of data for evaluating the steady-state electrical effects of 1100/1200- kV including AN, radio and television interference, electro- static effects, and effects on other utilities.

Transmission lines of 1100/1200-kV can be designed so electrical effects (i.e., maximum ground level electric field strength, AN, and oxidant production) are not signi- ficantly different from 500-kV lines. This requires the use of towers averaging 61 m in height and 8 subconductors for each phase. Each subconductor is 4.1 cm in diameter. A lateral profile of AN calculated for the 1100/1200 kV line is shown in Figure 7. The prototype line has been designed so that in two locations within a fenced study area, electric field strength could approach 14 kV/m. Other than possible damage to some trees which have purposely been left quite near the line, these higher field strengths are not expected

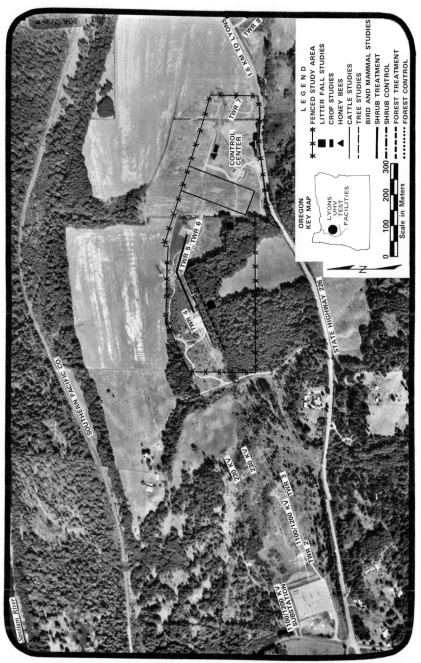

Figure 6 – The Bonneville Power Administration Lyons U.H.V.
Test Facilities showing location of biological
study areas.

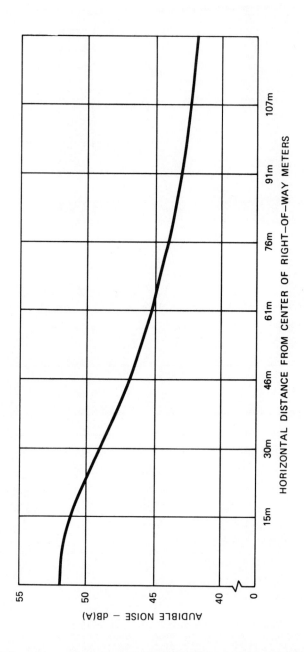

Figure 7 – Lateral profile for calculated mean audible noise
for the B.P.A. 1100/1200 kV Prototype Transmission
Line with eight 4.1 cm diameter conductors per
phase. Adapted from Biological Studies Task Team
(1977).

to result in any adverse biologic effects. By using these higher levels, however, the potential for threshold effects occuring just above design limits of 9 kV/m can be evaluated.

Biological studies at the 1100/1200-kV site include natural vegetation, crops, wildlife, cattle, and honey bees (Lee and Rogers 1976). From April to December 1976 pre-energization biological data was obtained with which to identify possible effects occurring after engerization. The general study approach after energization is to compare plants and animals on the 1100/1200-kV right-of-way with those in nearby control areas.

Total abundance and species diversity are the two main parameters which will be used to evaluate possible effects on bird populations. Observations of flight and feeding behavior will also be made. Birds inhabiting the study area are sampled along transects in four areas consisting of two treatments (right-of-way) and two controls (Figure 6). Birds seen, or heard within a radius of 20 m are counted during early morning surveys on three consecutive days.

Two hundred small mammal live traps are located along the same transects used for the bird studies. The traps are set for three consecutive nights and captured animals are marked with a unique toe clip pattern. Information on weight, approximate age, and reproductive condition is recorded each time an animal is captured.

Honey bees are being used as an indicator species to evaluate possible effects on insects. Honey bee studies take place during the spring and summer of 1977 and 1978. Six colonies are placed beneath the 1100/1200-kV line, and six away from the line (Figure 6). Parameters selected for study are honey and wax production, mortality, swarming tendencies, foraging, and general behavior. This phase of the study will complement a similar honey bee study sponsored by the Electric Power Research Institute (EPRI) which is being conducted beneath a 765-kV transmission line (Kornberg 1976).

Results of the biological studies, along with results of economic and engineering studies will provide the basis for a decision on whether to adopt 1100/1200-kV transmission in the BPA system. At the time this paper was written only a small amount of post energization biological data had been collected at the 1100/1200-kV test site.

Preliminary analysis of pre-energization bird data shows that although more birds were observed in the control areas than in the treatment areas, the differences were not statistically significant (Rogers 1977). Analysis of post-energization bird and mammal data will be done when sufficient data has been accumulated. No obvious adverse effects on bird and mammal populations have as yet been identified which could be attributed to AN or other operational parameters of the 1100/1200-kV line. During this time maximum line voltage was 1100 kV. Beginning in the summer of 1977 the line will be operated at 1200 kV during certain intervals. Preliminary measurements made since energization of the three phase 1100/1200 kV prototype line indicate the amount of AN produced by the line is within $\pm$ 2 dB of the calculated values shown in Figure 7.

Although very little foul weather AN data has as yet been obtained for the 1100/1200 kV line, preliminary measurements point out again that line design e.g., number and diameter of conductors, and not just line voltage alone determines the AN characteristics of a transmission line.

Only on a few occasions have behavioral responses been observed in animals which appeared to be a response to the electrical parameters of the 1100/1200-kV line. Shortly after the first phase of the line was engerized an American kestrel (Falco sparverius) attempted to land on an energized conductor. The bird approached to within approximately 30 cm of the conductor and after a few attempts at landing it finally flew off. The bird later landed on one of the unenergized phases of the 1100/1200-kV line. When all three phases of the 1100/1200 kV line were first energized the five head of cattle in the test pen were lying down almost directly under the line. When the line was energized there was a moderate amount of AN and four of the cattle immediately rose to their feet. After a few minutes two of the cattle laid down again near the line and the others began grazing nearby (John Hedlund, personal communication).

No behavioral effects have been noted in the honey bee colonies under the 1100/1200 kV line which could be attributed to operation of the line.

More data will be required to properly assess the possible biologic effects of the 1100/1200 kV line. Biological studies at the prototype site will continue through August 1978.

HVDC Transmission Line Biological Study

Relatively few d-c transmission lines have been constructed throughout the world. One of the first and longest
d-c lines built is the Celilo-Sylmar ± 400 kV d-c line which
extends from The Dalles, Oregon to Los Angeles, California
(Figure 1). A literature review failed to identify any
biological studies which had been conducted on a d-c transmission line right-of-way. With planning underway for a
possible second d-c intertie line, BPA developed a program
to study the effects of the existing Celilo-Sylmar d-c line
(BSTT 1976). A 13-month long study began in June 1976. The
BPA sponsored study is being conducted by the junior author.
The d-c biological study includes natural vegetation, crops,
wildlife, and domestic animals.

Although at various times observations were made along
the entire Oregon portion of the d-c line, two primary areas
were selected for intensive study. One was located in the
Western juniper (Juniperus occidentalis) vegetation zone and
the other in the big sagebrush (Artemisia tridentata) zone.
Almost 90 percent of the Celilo-Sylmar d-c line in Oregon is
located within these zones. Field studies were conducted
from mid-June 1976 through January 1977, and from May through
mid-June 1977. Because data analysis was not complete as
this paper was written only preliminary results are presented
here. We feel this is warranted because to our knowledge this
was the first biological study of a d-c transmission line
conducted in the U.S., and the measurements of AN made during
the study are among the first to be reported for the Celilo-
Sylmar d-c line which were made over an extended period of
time.

Power is transmitted on the Celilo-Sylmar ± 400-kV d-c
transmission line via a negative and positive pole each of
which consists of two 4.56 cm diameter subconductors. The
subconductors are strung on steel self-supporting or guyed
aluminum towers which are typically 36 m tall and 12 m wide
at the crossarms. Ground clearance to the conductors varies
from approximately 25 m at the towers to approximately 13 m
at midspan depending on site characteristics. Tower spacing
averages 2.9/km. A single overhead ground wire runs the
length of the Oregon portion of the d-c line. A light duty
maintenance road parallels the line approximately 15 m from
the center of the right-of-way. Distribution of ground level
electric field strengths for the d-c line is generally as-
symetrical and varies widely due to the pronounced effect of
wind on space change distribution (Bracken et al., 1977). At
no time during the study did we notice any effects which in-

dicated we could perceive the d-c fields.

Relatively few AN measurements had been made for the Celilo-Sylmar d-c line when this study was initiated although AN was measured during studies conducted with the d-c test line at The Dalles, Oregon (Hill et al., 1977). At various times during this study measurements of AN were made on the right-of-way and in control areas. These were made with a Bruel and Kjaer Type 2204 sound level meter with a 25.4 mm condenser microphone. With this microphone the meter had a range of 15-140 dB(A). The meter was fitted with an octave filter set with 10 center frequencies from 31.5 Hz to 16000 Hz. All measurements were made with the meter mounted on a tripod with the microphone in a vertical direction (perpendicular to the conductors). The microphone was covered with a polyurethane sponge windscreen when measurements were made. AN levels measured on the d-c transmission line right-of-way were generally lower and more variable than that reported for a-c lines of comparable voltage. Increases in AN above ambient for the d-c line ranged from essentially 0 up to 20 dB(A) with most measurements in the 0-10 dB(A) range. The highest AN level measured on the d-c line right-of-way during the study was 38 dB(A) which is probably about the maximum expected.

For wildlife studies emphasis was on songbirds, raptors, small mammals, mule deer (<u>Odocoileus</u> <u>hemionus</u>), and pronghorn antelope (<u>Antilocapra</u> <u>americana</u>). For purposes of this paper, only preliminary data from the songbird studies will be considered in detail. Songbirds were sampled using a circular plot technique. All birds seen and heard from fixed points were counted during early morning hours at 0.4 km intervals along 3.6 km long transects. The distances of the birds from the observer were measured with a rangefinder which had an upper limit of 219 m. Distances to birds further than this were estimated. In addition, the position of each bird was noted as being in one of eight 45 degree sections centered on the observer. During the 1977 spring sampling period this method was modified. A bearing was taken on each bird so that the distance of each bird from the center of the d-c line right-of-way could be estimated. In addition the length of the transects in the sagebrush study area was lengthened to 5.6 km with 15 stations. In the sagebrush study area, one transect was located on the center of the d-c line right-of-way and another in a control area parallel to the right-of-way and approximately 800 m away. In the Western juniper study area, a transect was established along a road which paralleled the right-of-way and was 400 m away. This was in addition to the right-of-way and control transects. In each study

area songbird sampling was conducted on alternate days until
a total of four days was completed for each transect for each
of three sampling periods.

During the 1976 songbird sampling period, a possible
sampling bias introduced by the effects of transmission line
AN on the observer was identified. Subjectively, it seemed
that AN of certain levels and quality might function to:
(1) increase the ambiguity of species identification, (2)
mask detectability of bird calls, or (3) bias distance estimation to those birds which were only heard. Since 50-80
percent of birds detected on right-of-way and control transects were heard but not seen these sources of bias could
affect the indices to songbird abundance and distribution
obtained from the censuses.

Sagebrush habitat, where transmission line construction
effects on vegetative structure are the least obvious, present the best field situation for examining the possible
sources and magnitude of these biases. Even so there were
differences in vegetation on the right-of-way compared with
the control areas. In the control as compared to the right-
of-way, big sagebrush occurred with slightly greater frequency, and the mean height and percent of total cover was
slightly greater. On the right-of-way green rabbit brush
(Chrysothamnus nauseosus) and grasses accounted for a significantly greater (P < 0.05) proportion of total cover than on
the control.

In 1976 and 1977 the total number of birds detected on
the right-of-way transects was respectively 76.2 percent and
75.2 percent of the total number detected on the control
transects. Unknown species of birds were not counted during
1976. In 1977 unkown species accounted for 13.7 percent of
the total number of birds detected on the right-of-way
transects and 12.4 percent of the total number of birds detected on the control transects. This difference of 1.3
percent was not statistically significant at the 0.05 probability level.

In 1977 the distances to birds heard but not seen were
classed as close, medium or far. These classes corresponded
to distances of less than 100 m, 100-200 m, and greater than
200 m, respectively. The four most abundant species of
songbirds were tabulated, considering only those birds heard,
into these classes (Table 4). The species were arranged in
a subjective ranking of song volume with the sage thrasher
the loudest and the horned lark the quietest.

Table 4.   Percentage Distance Distribution of Birds Heard
Only, On Control and Right-of-Way (ROW) Transects
in Sagebrush Habitat Along the $\pm$ 400 kV d-c Transmission Line in Central Oregon

| Species | Transect | Distance | | |
		Close 1/	Medium 2/	Far 3/
Sage Thrasher	Control	16.8	57.1	26.1
(Oreoscoptes	ROW	14.4	57.7	27.8
montanus)				
Sage Sparrow	Control	42.9	55.7	1.4
(Amphispiza	ROW	35.4	60.8	3.8
belli)				
Brewer's Sparrow	Control	72.3	27.7	0.0
(Spizella breweri)	ROW	61.1	38.9	0.0
Horned Lark	Control	87.0	13.0	0.0
(Eremophila	ROW	91.7	8.3	0.0
alpestris)				

1/   Less than 100 m
2/   100-200 m
3/   Greater than 200 m

There was little difference between right-of-way and
control distributions (Table 4) for the loud sage thrasher.
For the sage sparrow and Brewer's sparrow respectively, approximately 5 percent and 11 percent more birds were classed
in the medium distance category for the right-of-way transects
compared to the control. This could indicate that AN from
the transmission line was biasing distance estimates.
Analysis of the distance distribution from the transect
centerlines for birds seen, however, showed that the mean
distance was slightly greater, though not statistically
significant, for both sage sparrows and Brewer's sparrows
on the right-of-way compared to the control. The increased
frequency of these birds in medium and far categories may
reflect actual distribution patterns rather than AN bias on
the observer.

Horned larks present a somewhat different case. Their call and song is weak in comparison to the other birds in Table 4. It seems that the approximately 5 percent lower frequency of horned larks classed in the medium distance category on the right-of-way may reflect masking of this species' call by certain levels and quantities of AN from the transmission line.

If the AN from the transmission line markedly affected the total number of birds detected on the right-of-way we might expect the percent of total birds that are heard but not seen to be greater on the control transects as compared to the right-of-way. Some ambiguity in this relationship might be introduced by the fact that a number of the birds seen are first detected by auditory means.

In 1976, 79.9 percent of the total birds detected on the right-of-way were heard only and 71.5 percent of the total birds detected on the control transects were heard only. In 1977, 67.0 percent of birds detected on the right-of-way transects were heard only and 67.6 percent of birds detected on the control transects were heard only. In neither year was the difference between percent heard only on right-of-way and control transects significantly different.

The bias of AN on songbird detection on the $\pm$ 400 kV d-c transmission line right-of-way does not then appear to be of great enough magnitude to account for the approximately 25 percent lower number of birds detected on the right-of-way transects. Positive identification of this bias may have been obscured by the variable character of the AN from the d-c line. At each right-of-way transect station in 1977 the AN from the d-c line was classed into one of four categories: (1) no line AN perceived, (2) line AN perceived but of minor interference in songbird detection, (3) line AN perceived and of moderate interference in songbird detection, and (4) line AN perceived and of major interference in songbird detection. For the 60 transect stations (4 daily transects with 15 stations each day) the distribution of the AN into the above four subjective categories was:

category 1, 17 stations
category 2, 12 stations
category 3, 17 stations
category 4, 14 stations

We feel that AN bias becomes a more important consideration at higher transmission line AN levels than measured for the d-c line. The bias may be more subject to identification on a-c lines where AN is of generally higher levels and more consistent.

Wildlife Observations Near Two 500-kV Transmission Lines

As described above, during the study of the d-c transmission line we identified possible effects of AN that may
affect an observer's ability to detect songbirds on the right-
of-way. The possibility that AN may also affect songbird distribution was also raised.

To obtain additional information on these possibilities,
the senior author made observations along the right-of-way
of two 500-kV a-c transmission lines in south central Oregon.
These lines, Grizzly-Malin, run generally north and south and
in places are near to the Celilo-Sylmar d-c line (Figure 1).
These lines were constructed and energized during the late
1960's. The lines consist of steel "delta" towers and
each phase is strung with two 4.07 cm diameter subconductors.
Both lines have two overhead ground wires for lightning protection. An access road runs between the two lines. The
two parallel sets of towers are approximately 50 m apart.

The observation period (June 5-June 8, 1977) was timed
to coincide with similar field observations being conducted
on the d-c line. Two 3.6 km long sample transects were established. One was along the center of the right-of-way and
the other was 525 m west and parallel to the lines. The
study area was approximately 129 km south of Grizzly Substation (Figure 1). Topography of the study area was fairly
level and was in the big sagebrush vegetation type. We
chose the particular site because it was one of the few areas
along the right-of-way located completely in sagebrush. We
felt that the possible effects of AN would be more apparent
in such an area where relatively little difference in vegetation existed between the right-of-way and adjacent areas.

Sample methods employed were similar to those utilized
in the d-c transmission study. In addition to birds counted
at 10 stops along the transects, birds seen and heard between stops were also counted. The stops provided an opportunity to observe bird behavior under differing AN levels.

The location of each bird seen or heard was recorded by
noting both the angle and distance of the bird from the observer. A range finder with an upper limit of 180 m was
used to measure distances. A metal plate marked in degrees
and fitted to a tripod on which was mounted a 10x spotting
scope was used to determine bearings.

Counts began within one-half hour after sunrise. One
transect was usually completed by about 0900 hours and then

counts were made on the other transect while returning to
the starting point. The second transect was usually completed by about noon. Counts were conducted on four consecutive days beginning June 5. The transects were alter-
nated so that two early and two late morning counts were conducted for each transect.

AN measurements were made each day at about sunrise
before the counts began and then again after both transects
had been completed. At both times measurements were made on
and off the right-of-way. Measurements were made with the
Bruel and Kjaer type 2204 sound level meter which was described above. The meter was tripod mounted which put the
microphone about 1.5 m above the gound plane. Measurements
on the right-of-way were all made at the same point which
was mid span between two particular towers. The meter was
placed directly under conductors and the distance from the
microphone to the conductors was 11 m.

Figure 8 shows how AN levels varied between the right-of-
way and in the control area during the sample period. During
fair weather AN measured on the right-of-way averaged approximately 15 dB(A) higher than in the control area. With
wet conductors there was about a 20 dB(A) difference between
the two areas. The levels are for periods of minimal wind
and/or bird calls. We feel this best depicts the AN produced
by the lines in comparison to the relatively quiet surroundings of the study area. Wind or birds would at times
produce sound level meter readings in the control area which
equaled or exceeded readings on the right-of-way.

Although most AN measurements were made only on the
right-of-way or in the control area, on the night of June 4
measurements were made at two lateral distances near the
500-kV lines. With AN on the right-of-way at 46 dB(A) and
19 dB(A) in the control, 40 dB(A) was measured 25 m west of
the outermost conductors, and 36 dB(A) at 50 m.

During the sample period the 500-kV lines were always
audible on the right-of-way transect and frequently from the
control even during fair weather. This is in contrast to the
more intermittent AN from the d-c line.

Although some birds were detected at distances up to
200 m or more from the center of the right-of-way, AN from
the two 500-kV lines was sufficiently loud to cause some
ambiguity in the ability of the observer to detect birds by
their calls. The problems did not appear to be so much re-

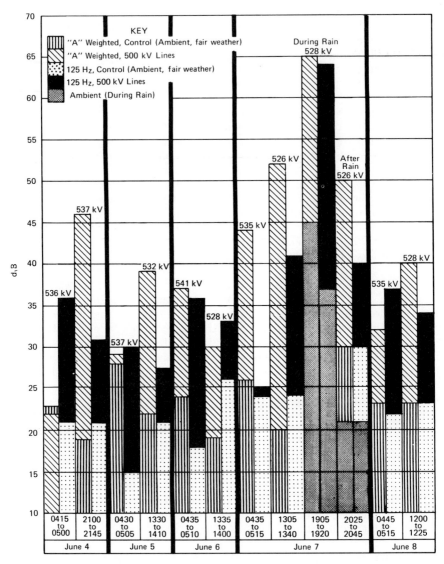

Figure 8 - Audible noise measured during June 1977 on the
right-of-way of two 500 kV a-c transmission lines
(Grizzly-Malin) as compared with a control area
525 m away from the line. During rain ambient
noise was also measured 1.6 km away from the line.
For the times shown above each day, the dB value
at the top of each keyed bar pattern represents
the average noise level measured during an approximate 3 minute period with essentially no wind. The
line voltage during the measurement period is shown
indicated in kV.

lated to hearing and locating birds by their songs but short, weak calls (e.g., cheeps, tweets) were sometimes difficult to locate when AN was originating from lines on either side of the observer. This effect, first noted on the d-c line, was more apparent with higher AN of the a-c lines.

In an attempt to eliminate some of the bias caused by the AN, we have tabulated only birds seen and heard within 75 m on either side of the right-of-way. In the field, it appeared the AN was not significantly affecting the observers ability to detect birds within this distance. Figure 9 shows the distribution of birds seen and heard during the four sample days within the 150 m wide transect on the right-of-way and in the same size transect in the control. As determined by a t-test for two sample means there was no significant difference in the mean distance of detected birds from the center of the right-of-way transect compared with the control. The relatively high number of birds detected near the center of the transects is in part due to birds being flushed by the observer.

Another explanation for the large number of birds observed in the center of the right-of-way may be possible effects of the access road. Birds along the road may have been more easily detected visually. On several occasions birds appeared to be attracted to the edge of the road. Even though the road was only 6 m wide the "edge effect" may have been a factor contributing to the greater number of birds near the center of the right-of-way. Additional information to evaluate this possibility could be obtained by conducting a count along a road near the transmission line which is of a similar width and which receives similar travel as the right-of-way. During the study period no vehicles were seen using the access road. On the weekend prior to the study a number of motorcyclists were observed using the access road and other roads in the area.

Although the detected birds appeared to be similarly distributed within the two transects, there was a statistically significant difference ($P < 0.05$) in the total number of birds detected in the two areas. Table 5 gives the total birds detected on each day for both seen and heard birds. Table 6 lists the bird species counted.

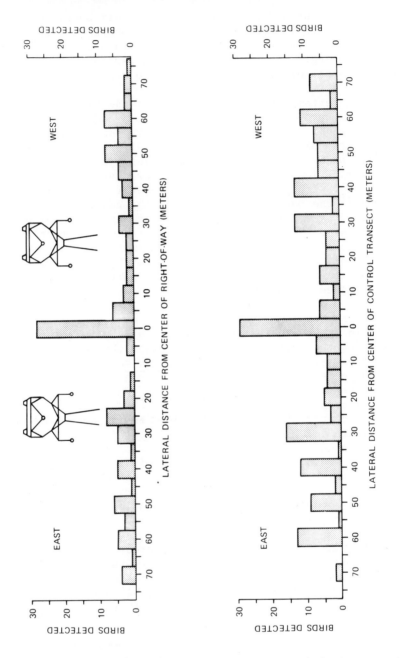

Figure 9 – Distribution of birds seen and heard during morning counts along a 150 m wide, 3.6 km long transect along the right-of-way of two 500 kV transmission lines and along a similar size control transect parallel to the right-of-way and 525 m west.

Figure 9 — Counts were made June 5–8, 1977 in south central
(continued) Oregon in Big Sagebrush (Artemisia tridentata)
vegetation type.

Table 5. The total number of birds seen and heard only, a-
long 3.6 km long, 150 m wide transects during
morning counts from June 5–8, 1977. One transect
was on the right-of-way of two 500-kV transmission
lines and the other was parallel to the lines and
525 m away.

Day	Control Transect			Right-of-way Transect		
	Birds			Birds		
	Seen	Heard only	Total	Seen	Heard only	Total
June 5	43	2	45 1/	19	5	24
June 6	28	23	51	25	8	33 1/
June 7	31	32	63 1/	22	6	28
June 8	19	20	39	29	11	40 1/
Totals	121	77	198	95	30	125

1/ These counts are for the early morning sample which began
one half hour after sunrise. The other counts were then
conducted from about 0900–1200 hours.

Table 6.   Species distribution of birds seen and heard along
3.6 km long, 150 m wide transects during morning
counts from June 5-8, 1977.   One transect was on
the right-of-way of two 500-kV transmission lines
and the other was parallel to the lines and 525 m
away.

Species 1/	Control		Right-of-way	
	Number Detected	Percent of Total	Number Detected	Percent of Total
Brewer's Sparrow (Spizella breweri)	97	49	46	36
Sage thrasher (Oreoscoptes montanus)	15	7.6	18	14.4
Sage sparrow (Amphispiza belli)	18	9.1	19	15.2
Horned lark (Eremophila alpestris)	6	3	0	0
Meadow lark (Sturnella neglecta)	0	0	1	0.8
Loggerhead shrike (Lanius ludovicianus)	0	0	4	3.2
American kestrel (Falco sparverius)	0	0	2	1.5
Mourning dove (Zenaidura macroura)	0	0	2	1.5
Unknown 2/	61	30.8	34	27.2
Total	198		125	

1/ The right-of-way also contained a raven nest with four
young on a tower.

2/ Most of the birds in this category were probably either
Brewer's sparrows or Sage sparrows.

Of particular interest in relation to AN bias is the ratio of birds that were heard only to the number seen which were not heard first. For the control transect this heard/seen ratio was 0.64 compared to 0.32 for the right-of-way. This indicates a noticeably greater proportion of auditory detections on the control transect. In addition, many birds that are eventually seen are first heard. AN then may also bias visual detections as well. On the control transect 44 percent of the birds seen were heard first and on the right-of-way the corresponding percentage was 35.

Assuming the AN did bias bird detectability, heard/seen ratios and the percent of seen birds that were heard first on the control, can be used to derive an estimate of the degree of bias present. We have calculated that in the present example, the bias introduced by AN could account for as much as 75 percent of the difference in total birds detected between the right-of-way and control transects. These calculations were based on the number of birds seen and not heard first on each transect. For these birds that were seen only, there was only 9 percent fewer detected in the right-of-way as compared to the control.

The magnitude of this bias would most likely vary among species due to differences in song volume and calling rates. The above calculations assume the same rate and intensity of bird vocalizations between the right-of-way and the control. AN could cause an increase in the frequency of calling rates similar to that reported by Potash (1972). This could function to lower the percentage of the difference in birds detected between right-of-way and control which could be accounted for by the AN bias.

Vegetation differences could in part be responsible for the remainder of the observed differences between right-of-way and control transects. Vegetation composition and distribution has consistently been shown to be an important factor which influences bird distribution (Tomoff 1974, Balda 1975). Although no quantitative measurements of vegetation on the right-of-way and control transects were made, in general the shrub density appeared to be very similar between the two areas. Noticeable exceptions included the bare, unsurfaced access road (approximately 6 m wide) on the right-of-way and areas near some towers which had apparently not completely revegetated since construction. This reduction in shrubs available for singing perches and feeding areas could cause some reduction in density.

Grue (1977:130) believed that removal of nesting and foraging habitat was a primary factor responsible for lower densities of breeding birds on the right-of-way of two 500-kV transmission lines in the desert shrub habitat in Arizona. Grue also counted birds heard and seen during his study although he did not believe AN from the lines biased his ability to detect birds by sound (Grue, personal communication).

Another possibility for the lower densities is that the transmission line itself including the electric field and/or audible noise caused some birds to avoid the right-of-way. Grue (1977) also suggested that these factors were at times responsible for lower bird densities on the right-of-way of two 500-kV lines in Arizona. On numerous occasions during the census counts, birds were observed perched and singing directly under the conductors during conditions of highly noticeable AN. Several other birds were observed flying across the right-of-way near the conductors. No unusual animal behavior was observed which obviously suggested an aversion to the right-of-way by birds which could be directly attributed to the AN or the electric field of the 500-kV lines.

An afternoon thunderstorm on June 7 provided an opportunity to make additional observations during a period when high levels of AN were being produced by the lines (Figure 8). Shortly after the rain began observations were made along the portion of the right-of-way containing the census transect. A golden eagle (<u>Aquila</u> <u>chrysaetos</u>) and a red-tailed hawk (<u>Buteo</u> <u>jamaicensis</u>) were observed perched on towers. The birds appeared to be seeking shelter from the storm. Although a golden eagle was seen flying from one of the towers the following morning, during the rain storm was the only other time during a 6-day stay near the lines that these species were seen on the towers.

There was a raven (<u>Corvus</u> <u>corax</u>) nest on one of the towers in the study area which contained four fully feathered young birds. While it was raining AN of 64 dB(A) and 69 dB for the 125 Hz frequency was measured at a point on the ground midway between the tower with the nest and the adjacent tower.

AN on the right-of-way after the rain measured between 52-56 dB(A). The sound of the two lines was like loud rushing water. Birds began singing again when the rain stopped. "Bits and pieces" of bird songs could be heard through the noise. In the only span checked, 6 birds, 2 jack rabbits and fresh (less than one hour old) coyote tracks, were seen all within 40 m from the center of the right-of-way. One of the birds, a sage thrasher, was perched and singing just beyond an outer conductor. The noise was so loud that the bird could just barely be heard from less than 60 m away. These observations of wildlife utilization of the right-of-way under high AN conditions shows that the AN was not causing birds and some other species to completely avoid the right-of-way.

We feel that the difference in numbers of birds detected on the right-of-way of the two 500-kV transmission lines as compared to the control can largely be explained by the negative bias of AN on the observer. The degree of this bias is probably partly determined by the hearing acuity of the observer. However, more intensive studies are needed to obtain more definitive information with which to evaluate our preliminary findings. The possibility that AN or the electric or magnetic field or other factors were affecting birds on and near the right-of-way to some degree cannot be ruled out. One possibility for obtaining more definitive information on the bias due to AN would be for the observer to wear ear plugs when conducting counts on both right-of-way and control transects. One problem, however, is that relatively few birds are detected solely by visual means. By completely eliminating hearing as a means of detecting birds the number of birds sampled would be extremely low. The effects of low sample sizes would then add to the difficulty of testing for differences between the right-of-way and control areas. To determine what levels of transmission AN mask bird calls and songs of various birds at varying distances from an observer, tape recorded sounds could be used in both laboratory and field situations.

Transmission Line Raptor Study

Preliminary work with raptors during the d-c transmission line study led to the development of a project to obtain information on the effects of other BPA transmission lines on this group of birds (Lee 1976). Until recent years, the effects, both beneficial and adverse, that transmission lines have on raptorial birds have not received a great deal of study.

Part of the BPA raptor study involves obtaining information with which to determine how extensively transmission line structures are utilized as raptor nesting sites. For purposes of this study raven nests are also being counted. Information is being obtained in an attempt to answer such

questions as: 1) which bird species are nesting on transmission structures? 2) where are nests located on various
type structures? 3) what site characteristics determine
whether transmission line structures are utilized for nesting
sites? 4) does the electrical and acoustical environment
within transmission towers have an effect on nesting birds?
5) how much annual raptor production occurs on BPA transmission line structures?

Birds nesting on EHV towers are exposed to AN and
electric and magnetic field levels which greatly exceed
those found at ground level. This provides an opportunity
to study the possible effects of these parameters on adult
birds and their young.

BPA is also installing a small number of artifical
raptor nest platforms to evaluate their effectiveness in
reducing maintenance problems caused by bird nests located
over conductors. The feasibility of providing more desirable nest sites for large raptors is also being studied.
The nest platform being installed was designed by Morlan
Nelson and the Idaho Power Company (Nelson and Nelson 1976).

On April 1. 1977 helicopter patrol observers began
collecting information on nesting raptors. Table 7 shows
the preliminary data collected for EHV lines through May
1977. It appears that many of these nests are successfully
producing young birds although this information is still being
developed. A much larger number of nests are located on
230-kV and lower voltage BPA transmission lines and this
information is also still being compiled.

Many of the nests on 500-kV lines are located on delta
configuration towers. Figure 10 shows a delta tower and gives
electric field strength and AN levels in some of the locations
where birds are known to perch and nest. It should be pointed
out that electric field measurements were made with a meter
designed to measure in uniform field areas such as on the
ground beneath a transmission line. The electric field in
the tower is actually perturbed by the metal structure so
the measurements in Figure 9 are only approximations. AN
levels were calculated by Mr. Vern Chartier of BPA based on
the work of Perry (1972) and on his own work with 765-kV
lines.

As shown in Figure 10 during rain AN within the 500-kV
tower strung with a single 6.35 cm conductor for each phase
could reach 76 dB(A) or more. Most 500-kV lines in the BPA
system utilize either the double or triple conductor bundle.

Table 7. Number of bird nests observed on BPA 500-kV a-c
          and ± 400-kV d-c transmission line structures
          during helicopter patrol flights in April and May
          1977.

	500 kV a.c.		± 400 kV d.c.	
	Forested*	Non-Forested+	Forested**	Non-Forested++
Total Length of Lines Patrolled (km)	2024	1552	103	323
Number of Nests:				
Hawks	5	41	6	4
Ravens	0	35	1	
Golden Eagle	0	0	0	1
Unknown	0	56		1
Total Nests	5	132	7	6

*   Lines in this category are located primarily in central
    and western Oregon and Washington and in Idaho in coni-
    ferous forest, vegetation types.
+   Lines in this category are located primarily in eastern
    Oregon and Washington in shrub/grass and grass vegetation
    types.
** This category consists primarily of Western juniper
    (Juniperus occidentalis) vegetation type.
++ This category consists primarily of sagebrush (Artemisia
    spp.) grass and cropland.

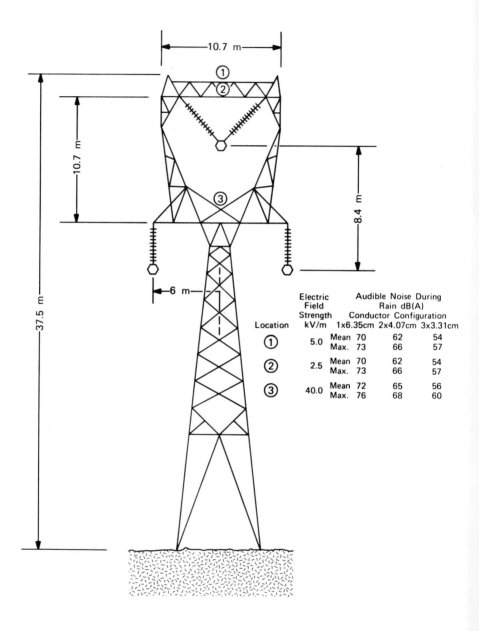

Figure 10 – Measured electric field strength and calculated audible noise for locations within a typical B.P.A. 500 kV a-c transmission line tower where raptors and other birds are known to perch and nest.

On one line that does have the single conductor design for its entire length (Vantage-Raver) a total of 13 nests were counted this spring. Apparently birds are not completely avoiding such lines because of their high AN characteristics.

It should be pointed out, however, that most of the nests on the Vantage-Raver 500-kV line are located east of the Cascade Mountains in a relatively dry climate. From March through May the mean precipitation is 51 mm and it is only 25 mm to 51 mm from June through August (Highsmith and Bard 1973:49). In this area precipitation and, therefore, the highest AN levels would occur only infrequently during the nesting season. The possible effects of AN of 76 dB(A) or more occurring for long periods of time on adult and young birds nesting in towers has yet to be determined. Woolf et al. (1976) reported that auditory stimulation of Japanese Quail eggs affected the developing embryo. It was found that 37 msec. bursts of sound of .0.1-8 kHz with a sound pressure level of 80 dB applied for two hours accelerated the time of hatching of the eggs by as much as 10 percent.

CONCLUSIONS

1. Transmission line audible noise (AN), produced by corona primarily from EHV lines, is one of many factors which contribute to the unique characteristics of a transmission line right-of-way. Any field study of the effects of transmisssion line AN on wildlife should acknowledge the possible synergistic actions of construction and maintenance activities, and effects due to electric and magnetic fields. To properly evaluate the potential impact of AN on wildlife requires a basic understanding of the technical aspects of AN.

2. The effect of transmission line AN on people is an environmental issue. Annoyance including possible interference with sleep is one of the most serious consequences of AN on people. When relating the human situation to wildlife several distinctions should be made. For wildlife, it is important to consider the maximum AN which occurs on the right-of-way. Differences in hearing abilities between man and wildlife should also be considered. With wildlife the effects of AN may not be entirely negative. It is possible that AN from transmission lines is a source of locational information for some species.

3.   Relatively few studies have been conducted to determine
      the effects of transmission lines on wildlife. Of those
      which have been done, still fewer have addressed the
      possible effects of AN or other electrical parameters.
      In studies conducted on Bonneville Power Administration
      transmission line rights-of-way both positive and
      negative effects of the right-of-way environment have
      been identified. The nature and magnitude of these
      effects has varied among species and taxonomic groups.
      Although we do not discount the possible effects of
      AN or electric fields, we believe that ecological effects
      observed to date can largely be explained by physical
      changes in habitat due to construction and maintenance
      activities.

4.   The effects of AN on even the most sensitive wildlife
      species may be difficult to detect in field studies due
      to marked temporal and spatial variation in wildlife
      populations. A possible exception is the reindeer, the
      behavior of which is reportedly affected by transmission
      line AN (Villmo 1972). The possibility of sampling bias
      due to the effects of AN on observers, and the fact that
      highest AN levels occur during foul weather, a time
      when wildife behavior is not well understood and when
      most wildlife observational techniques are severly im-
      paired add to the difficulties of identifying the
      effects of AN on wildlife.

5.   Increased public awareness of environmental pollutants
      such as transmission line AN has served to promote
      transmission line designs that minimize the production
      of AN. We would expect this trend to continue so that
      AN from newer EHV and UHV lines will be less than that
      produced by some of the first EHV lines constructed.

RECOMMENDATIONS

      Although a growing number of studies are being conducted
to determine the effects of transmission lines on wildlife,
few such studies acknowledge the unique characteristics of
the transmission line environment which distinguish it from
other rights-of-way. We recommend that persons conducting
such studies familiarize themselves with the technical characteristics of transmission lines. Most biologists have
little or no training which is applicable to the highly
specialized subject fields which are concerned with the
theory and operation of EHV and UHV (ultra-high-voltage)
transmission lines. We have found it necessary and extremely
helpful to consult with a number of persons who have the

necessary training to assist in designing and implementing
studies which consider the electrical parameters of transmission lines. At the minimum we believe that reports of
biological studies done on transmission line rights-of-way
should provide sufficient information to describe the total
right-of-way environment. Included should be such information
as the number and voltage of lines present, the range of
operating voltage during the study, the design of the line(s)
including the size and number of subconductors, and the
minimum conductor to ground clearances. Measurements of AN
or at least subjective judgements of the levels encountered
during the study should also be included. Other useful
information would include the dates when the line was constructed and energized, the use made of access roads, and
the kind of vegetation management activities utilized including information on the use of herbicides on the right-of-
way.

ACKNOWLEDGEMENTS

    Several persons provided much appreciated assistance
and advice during development and review of this paper. We
especially want to thank Dr. Dan Bracken, Physicist, and
Messrs. Steve Sarkinen, Stan Capon, and Vern Chartier,
Electrical Engineers, all from BPA, for their help. We
are also grateful to Dr. Lee Rogers and Mr. John Hedlund of
Battelle Northwest Laboratories for their review of the draft
and for use of preliminary data from their studies of the
BPA 1100/1200-kV Prototype Line.

    Funds for travel to Madrid for the senior author were
provided by the U.S. Environmental Protection Agency,
through the Office of Noise Abatement Control. We greatly
appreciate the support provided by that Agency. We especially
wish to thank Dr. John Fletcher for his efforts in arranging
the conference session on the effects of noise on wildlife
and for inviting our participation in this conference.

    This paper reports on research funded primarily by the
Bonneville Power Administration. BPA also provided financial
support and arrangements for travel.

REFERENCES CITED

1. Anderson, S.H., K. Mann, and H.M. Shugart, Jr. 1977.
   The Effect of Transmission Line Corridors on Bird
   Populations. The American Midland Naturalist. 97(1):
   216-221.
2. Anticaglia, J.R. 1970. Introduction: Noise in Our Over-
   polluted Environment. Pages 1-3 in Welch, B.L. and A.S.
   Welch (eds.) Physiological Effects of Noise. Plenum
   Press. New York. 365 pp.
3. Autrum, H. 1963. Anatomy and Physiology of Sound Re-
   ceptors in Invertebrates. Pages 412-433 in Busnel, R.G.
   (ed.) Acoustic Behavior of Animals. Elsevier Publishing
   Co. Amsterdam.
4. Balda, R.P. 1975. Vegetation Structure and Breeding
   Bird Diversity. Pages 59-80 in Smith, D.R. (Technical
   Coordinator) Proceedings of the Symposium on Management
   of Forest and Range Habitats for Nongame Birds. USDA
   Forest Service General Technical Report WO-1. Washington,
   D.C. 343 pp.
5. Biological Studies Task Team (BSTT). 1976. HVDC Transmission Biological Studies Program. Report on file in
   the Engineering and Construction Division Environmental
   Coordinator's Office. Bonneville Power Administration,
   Portland, Oregon.
6. Biological Studies Task Team (BSTT). 1977. Electrical
   and Biological Effects of Transmission Lines: A Review.
   Bonneville Power Administration. Portland, Oregon.
7. Bogert, Ch. M. 1960. The Influence of Sound on The
   Behavior of Amphibians and Reptiles in Lanyon, W.E.,
   and W.N. Tavolga (eds.) Animal Sounds and Communication.
   Publication No. 7 American Institute of Biological
   Sciences. Washington, D.C. 443 pp.
8. Bonneville Power Administration 1974. General Construction and Maintenance Program. Final Environment
   Statement. U.S. Dept. of the Interior, Bonneville
   Power Administration. Portland, Oregon 126 p.
9. Bonneville Power Administration. 1976a. 1,100/1,200 kilo-
   volt Transmission Line Prototype. U.S. Department of the
   Interior, Bonneville Power Administration, Portland,
   Oregon. 28 p.
10. Bonneville Power Administration. 1976b. 1100 kV Proto-
    type Final Facility Location Supplement Pages SAIS-1
    to 45 in Environmental Statement Fiscal Year 1976 Pro-
    posed Program, Facility Evaluation Appendix. Portland,
    Oregon.
11. Bracken, T.D., A.S. Capon, and D.V. Montgomery, 1977.
    Ground Level Electric Fields and Ion Currents on The

Celilo-Sylmar ± 400-kV D-C Intertie During Fair
Weather. Paper submitted for presentation at the IEEE
1977 Summer Power Meeting.

12. Bremond, J.C., 1963. Acoustic Behavior of Birds. Pages
709-750 in Busnel, R.G. (ed.). Acoustic Behavior of
Animals. Elsevier Publishing Co. Amsterdam.

13. Bridges, J.E. (Principal Investigator) 1975. Final Re-
port to Electric Power Research Institute for RP 381-1.
(2 vols.) I-- Biological Effects of High Voltage Electric
Fields: State-of-the-Art Review and Program Plan. II--
Bibliography onBiological Effects of Electric Fields.
IIT Research Institute. Chicago, Ill.

14. Buffington, J.D. 1974. Assessment of The Ecological
Consequences of Herbicide Use Along Transmission Line
Rights-of-way and Recommendation for Such Use. Argonne
National Laboratory, Argonne, Illinois. Available from
NTIS.

15. Busnel, R.G. and J. Giban. 1968. Prospective Considera-
tions Concerning Bio-Acoustics in Relation to Bird-
Scaring Techniques. Pages 17-18 in Murton, R.K. and
E.N. Wright (eds.) The Problems of Birds as Pests.
Academic Press, London and New York. 245 pp.

16. Canfield, R. H. 1941. Application of The Line Intercep-
tion Method in Sampling of Range Vegetation. Journal
of Forestry 39:388-394.

17. Carothers, S.W. and R.R. Johnson. 1975. Water Manage-
ment Practices and Their Effects on Nongame Birds in
Range Habitats. Pages 210-223, in Smith, D.R.
(Technical Coordinator) Proceedings of The Symposium
on Management of Forest and Range Habitats for Nongame
Birds. May 6-9, Tucson, Arizona. Forest Service U.S.D.A.
General Technical Report WO-1. Washington, D. C. 343 pp.

18. Cavanagh, J.B., D.P. Olson and S.N. Macrigeanis. 1976.
Wildlife Use of Power Line Rights-of-way in New Hamp-
shire. Pages 276-285, in Tillman, R. (ed.) Proceedings
of the First National Symposium on Environmental Concerns in Rights-of-way Management. Mississippi State
University, Mississippi. 335 pp.

19. Comber, M.G. and L.E. Zaffanella. 1975. Audible Noise.
Pages 192-247 in General Electric Company, Transmission
Line Reference Book 345 kV and Above. Electric Power
Research Institute, Palo Alto, California. 393 pp.

20. Comptroller General of the United States. 1977. Report
to the Congress Noise Pollution- Federal Program to
Control It Has Been Slow and Ineffective. U.S. Environmental Protection Agency and U.S. Department of Transportation. Washington, D. C. 63 pp.

21. Deno, D.W. and M.G. Comber, 1975. Corona Phenomena on AC Transmission Lines. Pages 122-148, in General Electric Company Transmission Line Reference Book 345-kV and Above. Electric Power Research Institute, Palo Alto, California. 393 p.
22. Deno, D.W. and L.E. Zaffanella. 1975. Electrostatic Effects of Overhead Transmission Lines and Stations. Pages 248-280 in General Electric Co., Transmission Line Reference Book 345-kV and Above. Electric Power Research Institute, Palo Alto, California. 393 pp.
23. Driscoll, D.A. 1975. Prepared Testimony (Effects of Audible Noise, Ozone, Induced Currents and Voltages, and Electric and Magnetic Fields From 765 kV Transmission Lines. Cases 26529 and 26599 New York Public Service Commission.
24. Edgerton, P.J. 1972. Big Game Use and Habitat Changes in a Recently Logged Mixed Conifer Forest in Northeastern Oregon. Proc. Western Assn., State Game and Fish Comm. 52:239-246.
25. Egler, F.E. 1953. Our Disregarded Rights-of-way - Ten Million Unused Wildlife Acres. Trans. N. Am. Wildl. Conf. 18:147-158.
26. Egler, F.E. 1957. Right-of-ways and Wildlife Habitat: A Progress Report. Trans. N. Am. Wildl. Conf. 22:133-144.
27. Emlen, J.T. Jr. 1960. Introduction. Pages ix-xiii in Lanyon, W.E. and W. N. Tavolga (eds.) Animal Sounds and Communication. Publication No. 7. American Institute of Biological Sciences. Washington 6, D.C. 443 pp.
28. Environmental Protection Agency (EPA) 1974. Information on Levels of Environmental Noise Requisite to Protect Public Health and Welfare with an Adequate Margin of Safety. Washington, D.C.
29. Environmental Protection Agency (EPA). 1972. Noise Pollution. Washington, D.C. 13 pp.
30. Evans, E.F. 1968. Cortical Representation. Pages 272-287, in de Reuch, A.V.S. and J. Knight (eds.). Hearing Mechanisms in Vertebrates. Ciba Foundation Symposium. Churchill, London.
31. Fletcher, J.L. 1975. Prepared Testimony (Effects of Noise on Wildlife and Domestic Animals). Common Record Hearings on Health and Safety of 765 kV Transmission Lines. Cases 26529 and 26599. New York Public Service Commission.
32. Frings, H. and M. Frings. 1959. Reactions of Swarms of Pentaneura aspera (Diptera: Tendipedidae) to Sound. Annals of The Entomological Society of America 52:728-733.
33. Frings, H. 1964. Sound in Vertebrate Pest Control. Vertebrate Pest Control Conference. 2:50-56.

34. Goodland, R. 1973. Powerlines and the Environment.
     The Cary Arboretum of the New York Botanical Gardens.
     Millbrook, New York. 170 pp.
35. Goodwin, J. G., Jr., 1975. Big Game Movement Near A
     500-kV Transmission Line in Northern Idaho. A Study
     by the Western Interstate Commission for Higher
     Education for the Engineering and Construction Divi-
     sion, Bonneville Power Administration. 56 pp.
36. Goodwin, J.G., Jr. and J.M. Lee Jr., 1975. A Study of
     500-kV Transmission Lines and Wildlife, and BPA
     Activities Related to Biologic Effects of Transmis-
     sion Lines. Paper Presented at the Meeting of the
     Utilities Appearance Committee in Portland, Oregon.
     September.
37. Granit, O. 1941. Beitrage zur Kenntnis des Gehorsinns
     des Vogel. Ornis Fennica. 18:49-71.
38. Griffin, D.R. 1976. The Audibility of Frog Choruses
     To Migrating Birds. Animal Behavior 24(2):421-427.
39. Grue, C.E. 1977. The Impact of Powerline Construction
     on Birds in Arizona. M.S. Thesis. Northern Arizona
     University, Flagstaff. 265 pp.
40. Haskell, P.T. and P. Belton. 1959. Electrical Re-
     sponses of Certain Lepidopterous Tympanal Organs.
     Nature. 177:139-140.
41. Highsmith, R.M. Jr. and R. Bard. 1973. Atlas of the
     Pacific Northwest 5th Edition. Oregon State Univer-
     sity Press. Corvallis. 128 pp.
42. Hill, H.L., A.S. Capon, O. Ratz, P.E. Renner, and W.
     D. Schmidt. 1977. Transmission Line Reference Book
     HVDC to $\pm$ 600 kV. Prepared by the Bonneville Power
     Administration and Published by Electric Power
     Research Institute, Palo Alto, California. 170 pp.
43. Ianna, F., G.L. Wilson, and D.J. Bosack. 1973.
     Spectral Characteristics of Acoustic Noise From
     Metallic Protrusions and Water Droplets in High
     Electric Fields. Paper presented at the IEEE Winter
     Meeting, New York, January 28-February 2.
44. IEEE Committee Report. 1972. A Guide For the Measure-
     ment of Audible Noise From Transmission Lines. IEEE
     Transactions Power Apparatus and Systems. Vol PAS-91:
     853-856.
45. Janes, D.E. 1976. Background Information on Extra
     High-Voltage Overhead Electric Transmission Lines.
     U.S. Environmental Protection Agency, Office of
     Radiation Programs, Electromagnetic Radiation Analysis
     Branch. Silver Spring, Maryland.
46. Jenkins, D.W. 1972. Development of a Continuing
     Program to Provide Indicators and Indices of Wildlife
     and the Natural Environment. Smithsonian Institution.
     Washington, D.C. 165 pp.

47.   Kitchings, J.T., H.H. Shugart, and J.D. Story. 1974.
       Environmental Impacts Associated with Electric
       Transmission Lines. Environmental Sciences Division.
       Oak Ridge National Laboratory, Oak Ridge, Tennessee.
       96 pp.

48.   Klein, D.R. 1971. Reaction of Reindeer to Obstruc-
       tions and Disturbances. Science 173(3995):393-397.

49.   Kolcio, N.B., J.Ware, R.L. Zagier, V.L. Chartier,
       and F.M. Dietrich. 1973. The Apple Grove 750 kV
       Project Statistical Analysis of Audible Noise Per-
       formances of Conductors at 775 kV. Paper T 73 437-1.
       Presented at the IEEE Summer Power Meeting. July 15-
       20. Vancouver B.C.

50.   Kornberg, H.A. 1976. EPRI's Research Program on Bio-
       logic Effects of Electric Fields. Pages 136-141 in
       Tillman, R. (ed.) Proceedings of the First National
       Symposium on Environmental Concerns in Rights-of-way
       Management. January 6-8, 1976. Mississippi State
       University, Mississippi.

51.   Krueger, A.P. and E.J. Reed, 1976. Biological Impact
       of Small Air Ions. Science. 193:1209-1213.

52.   Larking, R.P. and D.J. Sutherland. 1977. Migrating
       Birds Respond to Project Seafarer's Electromagnetic
       Field. Science. 195(4280):777-778.

53.   Lay, W.D. 1938. How Valuable are Woodland Clearings
       to Bird Life? Wilson Bulletin 50:254-256.

54.   Lee, J.M. Jr. 1976. A Study of the Effects of BPA
       Transmission Structures on Raptors and the Experi-
       mental Installation of Raptor Nesting Platforms. A
       Project Description and Environmental Analysis. Report
       on File in the Engineering and Construction Division
       Environmental Coordinator's Office. Bonneville Power
       Administration, Portland, Oregon.

55.   Lee, J.M. Jr. and L.E. Rogers. 1976. Biological
       Studies of a 1200-kV Prototype Transmission Line
       Near Lyons, Oregon. Quarterly Progress Report No. 1.
       Report Available from the E&C Environmental Coordina-
       tor's Office, Bonneville Power Administration, Port-
       land, Oregon.

56.   Lee, J.M. Jr. 1974. Preliminary Investigation of
       Elk Migration in Relation to Engineer Power Trans-
       mission Lines. Report on File in the Engineering
       and Construction Division Environmental Coordinator's
       Office, Bonneville Power Administration, Portland,
       Oregon. 8 p.

57.   Llaurado, J.G., A Sances, Jr., and J.H. Battocletti
       (eds.) 1974. Biologic and Clinical Effects of Low-
       Frequency Magnetic and Electric Fields. Charles C.
       Thomas, Springfield, Illinois. 345 pp.

58. Lott, J.R. and H. B. McCain. 1973. Some Effects of Continuous and Pulsating Electric Fields on Brain Wave Activity in Rats. Int. J. Biometeor. 17(3):221-225.

59. Martinka, B. 1974. The New Horizon. Montana Outdoors 5(4):16-21.

60. Meyer, T.D. 1976. An Evaluation of Chemically-Sprayed Electric Transmission Line Rights-of-way for Actual and Potential Wildlife Use. Pages 287-294, in Tillman, R. (ed.) Proceedings of the First National Symposium on Environmental Concerns in Rights-of-way Management. Mississippi State University. Department of Wildlife and Fisheries. Mississippi State.

61. Memphis State University. 1971. Effects of Noise on Wildlife and Other Animals. Report prepared for the U.S. Environmental Protection Agency. Contract 68-04-0024. 73 pp.

62. Miller, D.A. 1974. Electric and Magnetic Fields Produced by Commercial Power Systems. Pages 62-70 in Llaurado, J.G. et. al. (eds.) Biological and Clinical Effects of Low-Frequency Magnetic and Electric Fields. Charles C. Thomas, Springfield, Illinois. 345 pp.

63. Nelson, N.W. and P. Nelson. 1976. Power Lines and Birds of Prey. Idaho Wildlife Review. March-April, 1976. p. 3-7.

64. Norris, L.A. 1971. Herbicide Residues in Soil and Water from Bonneville Power Administration Transmission Line Rights-of-way. This and other reports in this ongoing study available from Bonneville Power Administration. Portland, Oregon.

65. Odum, E.P. 1959. Fundamentals of Ecology. W.B. Saunders Co. Philadelphia and London.

66. Overton, W.S. and D.E. Davis. 1969. Estimating the Numbers of Animals in Wildlife Populations. Pages 403-455, in Giles, R.H. Jr., (ed) Wildlife Management Techniques. The Wildlife Society. Washington, D.C. 623 pp.

67. Patton, D.R. 1974. Patch Cutting Increases Deer and Elk Use of a Pine Forest in Arizona. Journal of Forestry. 72(12):764-766.

68. Pengelly, W.L. 1973. Clearcutting and Wildlife. Montana Outdoors 4(6):26-30.

69. Perry. D.E. 1972. An Analysis of Transmission Line Audible Noise Levels Based Upon Field and Three-Phase Test Line Measurements. IEEE Transactions of Power Apparatus and Systems, Vol. PAS-91. p. 857.

70. Phillips, R.D., R.L. Richardson, W.T. Kahne, D.L. Hjeresen, J.L. Beamer, and M.F. Gillis. 1976. Effects of Electric Fields on Large Animals A Feasibility

Study. Final Report to Electric Power Research
Institute on Research Project RP 581-1. Battelle
Pacific Northwest Laboratories, Richland, Washington.
73 pp.

71.   Polk, C. 1974. Sources, Propagation, Amplitude,
and Temporal Variation of Extremely Low Frequency
(0-100 Hz) Electromagnetic Fields. Pages 21-48 in
Llaurado, J.G. et. al. (eds.) Biological and Clinical
Effects of Low-Frequency Magnetic and Electric Fields.
Charles C. Thomas, Springfield, Illinois 345 pp.

72.   Potash, L.M. 1972. A Signal Detection Problem and
Possible Solution in Japanese Quail (Coturnix
coturnix Japanica) Animal Behavior 20(1):192-195.

73.   Roach, J.F., F.M. Dietrich, V.L. Chartier, and H.J.
Nowak. 1977. Ozone Concentration Measurements on
the C-Line at the Apple Grove 750 kV Project and
Theoretical Estimates of Ozone Concentrations Near
765 kV Lines of Normal Design. Paper submitted for
presentation at the IEEE Summer Power Meeting.

74.   Roeder, K.D., and A.E. Treat. 1957. Ultrasonic
Reception by the Tympanic Organ of Noctuid Moths.
Journal of Experimental Zoology. 134:127-158.

75.   Rogers. L.E. 1977. Monthly Progress Report No. 12.
Biological Studies of an 1100-kV Prototype Transmission Line Near Lyons, Oregon. A Report to Bonneville Power Adminstration by Battelle Pacific Northwest Laboratories, Richland, Washington. 22 pp.

76.   Sales, G. and D. Pye. 1974. Ultrasonic Communication
by Animals. Chapman and Hall, Ltd. London. 281 pp.

77.   Schaller, F. and C. Timm. 1950. Das Horvermogen der
Nachtschmetterlinge. Z. Vergleich. Physiol. 32-287-
302.

78.   Schaller, F. and C. Timm. 1949. Schallreaktioneu
bei Nachtfalfern. Experientia. 5.162.

79.   Schreiber, R.K., W.C. Johnson, J.D. Story, C. Wenzel,
and J.T. Kitchings. 1976. Effects of Powerline
Rights-of-way on Small, Nongame Mammal Community
Structure. Pages 263-273, in Tillman, R. (ed.) Proceedings of the First National Symposium on Environmental Concerns in Rights-of-way Management. Mississippi State University. Department of Wildlife and
Fisheries. Mississippi State.

80.   Schwartzkopff, J. 1955. On the Hearing of Birds.
Auk. 72:340-347.

81.   Sebo, S.A., J.T. Heibel, M.Frydman, and C.H. Shih.
1976. Examination of Ozone Emanations From EHV
Transmission Line Corona Discharges. IEEE PAS 95 p.
693-703.

82. Sotavalta, O. 1963. The Flight Sounds of Insects. Pages 374-390 in Busnel, R.G. (ed.). Acoustic Behavior of Animals. Elsevier Publishing Co. Amsterdam.

83. Stahlecker, D.W. 1975. Impacts of a 230-kV Transmission Line on Great Plains Wildlife. M.S. Thesis. Colorado State University. For Collins. 67 pp.

84. State of New York Public Service Commission (SNYPSC). 1976. Opinion and Order Authorizing Erection of Support Structures and Conductors (765 kV). Opinion No. 76-12. Case 26529. Power Authority of the State of New York. Albany.

85. Stewart, J.L. 1974. Experiments with Sounds in Repelling Mammals. Proceedings of the Vertebrate Pest Control Conference 6:222-226.

86. Strother, W.F. 1959. The Electrical Response of the Auditory Mechanism in the Bullfrog (Rana catesbeiaua). Journal of Comparative Physiology and Psychology. 52:157-162.

87. Thomas, W.A., G. Goldstein, and W.H. Wilcox. 1973. Biological Indicators of Environmental Quality. A Bibliography of Abstracts. Ann Arbor Science Publishers, Inc. Ann Arbor, Mich. 254 pp.

88. Tillman, R. (ed.). 1976. Proceedings of the First National Symposium on Environmental Concerns in Rights-of-way Management. January 6-8, 1976. Mississippi State University, Mississippi. Available from D.H. Arner, Head Dept. Wildlife and Fisheries, P.O. Drawer LW.

89. Tischner, H. 1953. Uber den Gehorsinn von Stechmuckern. Acustica. 3:335-343.

90. Tomoff, C.S. 1974. Avian Species Diversity in Desert Scrub. Ecology 55(2):396-403.

91. Trainer, J.R. 1946. The Auditory Acuity of Certain Birds. Cornell University Abstracts of Theses. Pages 246-251.

92. Traux, C.J. 1975. EHV-UHV Transmission Systems. Pages 11-45 in General Electric Company, Transmission Line Reference Book 345 kV and Above. Electric Power Research Institute. Palo Alto, California. 393 pp.

93. United States Department of the Interior (USDI) and United States Department of Agriculture (USDA). 1970. Environmental Criteria for Electric Transmission Systems. Washington, D.C. 52 pp.

94. Villmo, L. 1972. The Scandinavia Viewpoint. Pages 4-9 in Luick, J.R., P.C. Lent, D.R. Klein, and R.G. White (eds.). Proceedings of the First International Reindeer and Caribou Symposium. Biological Papers of the University of Alaska Special Report No. 1. Fairbanks.

95.   Walmo, O.C., W.L. Regelin, and D.W. Reschert. 1972.
      Forage Use by Mule Deer Relative to Logging in
      Colorado. Journal of Wildlife Management. 36(4):1025-
      1033.

96.   Wever, E.G. and C.H.W. Bray. 1933. A New Method for
      the Study of Hearing in Insects. Journal of Cellular
      Comparative Physiology. 4:79-93.

97.   Wever, E.G., and J.A. Vernon. 1957. The Auditory
      Sensitivity of the Atlantic Grasshopper. Proceedings
      of the National Academy of Science. U.S. 43:346-348.

98.   Woolf, N.K., J.L. Bixby, and R.R. Capranica. 1976.
      Prenatal Experience and Avian Development: Brief
      Auditory Stimulation Accelerates the Hatching of
      Japanese Quail. Science. 194(4268):959-960.

99.   Yodlowski, M., M.L. Kreithen, and W.T. Keeton. 1977.
      Detection of Atmospheric Infrasound by Homing Pigeons.
      Nature 265:725-726.

100. Young, L.B., and H.P. Young. 1974. Pollution by
      Electrical Transmission. Bulletin of the Atomic
      Scientists. 30(10):34-38.

OCEAN NOISE AND THE BEHAVIOR OF MARINE ANIMALS:
RELATIONSHIPS AND IMPLICATIONS

Arthur A. Myrberg, Jr.

Rosenstiel School of Marine and Atmospheric Science
University of Miami
Miami, Florida

INTRODUCTION

There is general agreement among biologists that the
acoustical sense of aquatic animals probably constitutes
their most important distance receptor-system. This is particularly true among members of two vertebrate lines, the
fishes and the marine mammals, whose acoustical activities
have been investigated at an ever-increasing rate of sophistication during recent years. A major conclusion drawn from
these varied studies - inclusive of those using psycho-
physical, physiological and ethological methodologies - is
that the acoustical system can, and does, provide its owner
appropriate information, readily and rapidly, on a variety of
functions relative to food, competitors, potential mates and
predators.

Concomitant with the increasing sophistication of pro-
blem-queries, experimental designs and available instrumentation, there is a growing awareness that ambient noise, itself,
can no longer be ignored in underwater bioacoustics. Not
only is it probable that such noise actually affects, at
least temporarily, the hearing abilities of the animals concerned, it may also act to inhibit sound production as well.
Such a reduction in acoustic transmission and/or its reception
can, of course, adversely affect the reproductive potential
or even the survival of any given species or population that
is dependent on such a sensory process. Additionally,
evidence is beginning to show up which indicates that excessive noise can also have other more direct, deleterious consequences on marine biological systems (see below). This
problem-area will become more widely recognized as more is

learned about ambient noise, especially in the shallow, coastal regions of the world's oceans.

The present state of our rather limited knowledge, regarding acoustical noise and its effects on selected marine biological systems, will be reviewed below. Also, I shall take this opportunity to speculate, hopefully in not an undue fashion, regarding how such knowledge may provide insight into problems that extend beyond those restricted to bioacoustics. The scope of review has been limited to relevant findings from marine fishes and selected marine mammals, specifically the odontocete cetaceans and pinnipeds, since little or nothing is presently known about the subject in other groups. Also, only the shallow, coastal regions of the Continental Shelf are considered since they represent the major habitats of those animals considered here.

SHALLOW-WATER AMBIENT NOISE

Measurements of shallow-water, ambient noise from widely differing locations (e.g., Arase and Arase, 1966; Banner, 1968, 1972; Dietz et al., 1960; Heindsmann et al., 1955; Myrberg et al., 1972, 1976; Piggott, 1964; Tavolga , 1974a; Wenz, 1962, 1972; Widener, 1967) have shown that sound-levels in coastal waters - including bays and harbors - vary greatly in time and place (Albers, 1965; Urick, 1975; Wenz, 1964). Also, at frequencies below 1000 Hz, such noise often constitutes a mixture of different types of noise (Fig. 1 - following page), each originating from different sources, e.g., traffic (and industry), wind, and soniferous marine animals. At frequencies higher than 1000 Hz, still another type, rain noise, replaces the predominantly low frequency traffic (or industrial) noise as a contributing factor. Despite the considerable variation in existing noise at given locations from time to time, if conditions remain stable for reasonable periods, the resulting curves do show some predictive similarity and constancy.

Figure 1 provides a general synthesis of the ambient noise conditions that prevail for shallow, coastal regions whose depths are generally less than 70 m. The levels attributed to respective sea states have been taken from Piggott (1964), their slopes being similar to those determined off the coast of Bimini, Bahamas, at comparable states and recording depth (see Myrberg et al., 1969). Such levels, being strongly dependent upon the wind (even to 10 Hz - Piggott, 1964), are generally 5 to 7 dB higher than corresponding levels for the same sea state in deep waters (Knudsen et al., 1948; Wenz, 1962), especially at frequencies

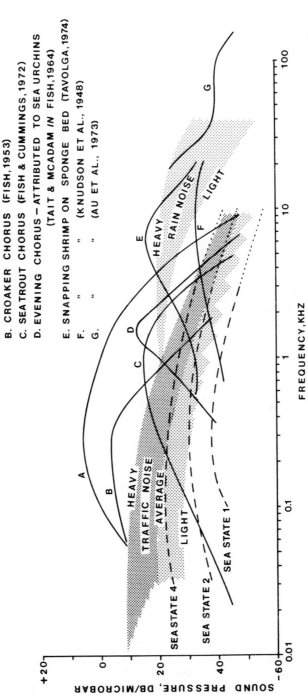

EXAMPLES OF BIOLOGICAL SOURCES OF
SUSTAINED AMBIENT NOISE

A. CROAKER CHORUS (FISH, 1953)
B. CROAKER CHORUS (FISH, 1953)
C. SEATROUT CHORUS (FISH & CUMMINGS, 1972)
D. EVENING CHORUS — ATTRIBUTED TO SEA URCHINS
   (TAIT & MCADAM IN FISH, 1964)
E. SNAPPING SHRIMP ON SPONGE BED (TAVOLGA, 1974)
F.   "   "   "   (KNUDSON ET AL., 1948)
G.   "   "   "   (AU ET AL., 1973)

Figure 1 – Shallow water (<70 m), ambient noise (spectrum level). Data have been extracted from numerous sources and redrawn. Major references include Albers, 1965; Urick, 1975 (traffic noise); Piggot, 1964 (sea state); and Heindsmann et al, 1955 (rain noise).

171

above 500 Hz.   Also below that point, the corresponding de-
flection of the curves representing specific sea states are
not parallel.

Corresponding levels for traffic (and industrial) noise
across the spectrum were taken from Albers (1965) and Urick
(1975) with the relatively flat, high-level, low-frequency
extension between 10 and 50 Hz taken from Myrberg et al.
(1976).   The "nature" of the curve in this particular region
has been recently reviewed by Wenz (1972).   The slopes of
these respective curves are in reasonable agreement with
those obtained by Maniwa (1971) and Olsen (1971) from
noise measurements made in the vicinity of large fishing
vessels.   There can be little doubt that under quiet or
reasonable sea states, traffic and/or industrial activity
may constitute, in certain locations, a major source of noise
in the two decades, between 10 and 1000 Hz (see Urick, 1975,
for an insightful review).

Levels of rain noise were provided from the date supplied by Heindsmann et al., (1955).   Within the spectral region of influence, there is, likewise, no question that rain
is a significant source of background noise.

The most difficult type of ambient noise to characterize by predictions through time and space is that produced
by soniferous animals.   Although often intermittent, it can,
at times, characterize a given region.   Biological sounds in
the sea are many and varied; a few examples are provided in
Fig. 1.   The latter were chosen since their levels were
consistently high (as much as 50 dB above normal) for long
periods of time.   Kaneohe Bay, Hawaii and Bimini Bay, Bahamas
(E and G, respectively) are examples of specific areas that
are characterized by high-level, ambient-noise above 2000 Hz,
this resulting from enormous populations of snapping shrimp.
Such sounds, be they from fish , shrimp, sea urchins or
otherwise, often span a broad frequency range, with peaks of
intensity being dependent on the animals involved.   Finally,
many such sources are associated with particular seasons of
the year, phases of the moon, or times of the day (or night)
(for review, see Cummings et al., 1964 and Steinberg et al.,
1965).

SOUND DETECTION AND LOCALIZATION BY FISHES AND MARINE MAMMALS

Recent reviews of our knowledge concerning the detection and localization of sounds by fishes (Chapman, 1973;
Fay, 1974; Hawkins and Chapman, 1975; Hawkins, 1974 ; Popper
and Fay, 1973; Tavolga, 1971, 1974b ; Sand and Enger, 1974,
Schuijf, 1974) preclude the necessity of reviewing this broad
subject here.   I shall, therefore, only briefly discuss a few

topics that appear relevant to the subject at hand, i.e., ambient noise. This will also be the case regarding the specific mammals of interest - the odontocete (toothed) cetaceans and pinnipeds (for recent reviews, see Evans, 1973, Norris, 1969 and Schusterman et al., 1972).

Selected audiograms of three marine fishes and three marine mammals are provided in Figure 2, the examples chosen in each group possessing different levels of sensitivity. These are provided for direct comparison of thresholds relative to spectral range and sensitivity. An aerial audiogram for man is also included to show that directly comparable levels of sensitivity are attained by at least some species of fishes at lower frequencies and by various species of marine mammals at a higher spectral range.

FISHES

Hearing studies, carried out on many species found in widely differing habitats, show that although certain marine fishes can detect quite high frequencies (e.g., herrings - 5 to 10 KHz), the great majority are sensitive only to frequencies below 2000 Hz. The major exception to this "rule", the cypriniform fishes (e.g., minnows, carp, goldfish, catfishes), though also detecting frequencies up to between 5 and 10 KHz, are poorly represented in the marine environment. They are, therefore, not discussed here.

It is difficult to form meaningful generalizations from the audiograms that are presently available for more than thirty species of marine fishes (many are provided in the review by Popper and Fay, 1973). There are, nevertheless, certain patterns of relative consistency. Besides the dichotomy that is often made between the auditory "specialists" (i.e., the cypriniform fishes and other selected species) and the so-called "non-specialists" (i.e., those showing a reduced range of spectral sensitivity along with a general other group), the latter can also be divided into two groups, based largely on speculative candor and a desire for convenience. The members of Group 1 (See Figure 3) have a rather broad region of greatest sensitivity at frequencies generally between 75 and 300 Hz, while those of Group 2 (Figure 4) have a narrower range of greatest sensitivity at frequencies around 400 to 800 Hz. Also, although there are exceptions, the ascending slopes of decreasing sensitivity above the regions of peak sensitivity are rather steep (35 to 40 dB/octave) and similar among many species from both groups, regardless of their relationships, one with another. This

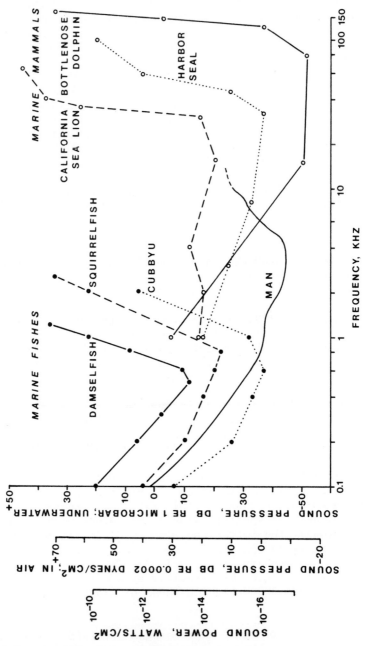

Figure 2 — A comparison of hearing curves among selected marine fishes, marine mammals (authors cited in text), and man (aerial curve — Sivian and White, 1933).

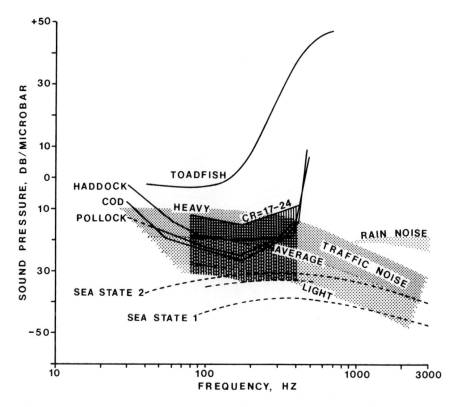

Figure 3 - Low frequency ambient noise and its probable masking effect on the hearing abilities of selected marine fishes (Group 1 - see text), whose peak sensitivities are found within that spectrum. The four audiograms shown were determined either totally, or partially, in the field. The hatched area is the region chosen to show the amount of masking that would occur above the arbitrarily chosen spectrum levels of sea state (< 2) and traffic noise (light, see Fig. 1), for those species possessing the critical ratios (CR), as given. Note the great similarity between the slopes of the simulated threshold and those of the actual thresholds of the cod and pollock that were obtained by Chapman (1973) under quieter conditions (see text for further information).

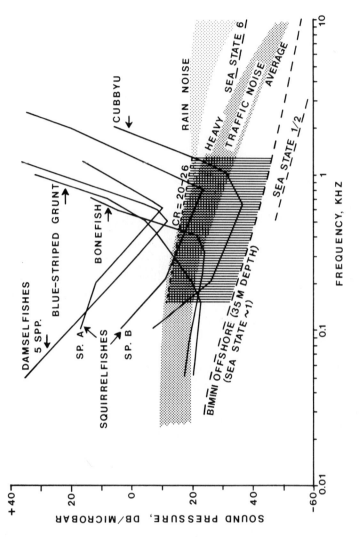

Figure 4 - Low frequency ambient noise and its probable masking effect on the hearing abilities of selected marine fishes (Group 2 - see text), whose peak sensitivities are found within the spectrum of interest. The hatched area is the region chosen to show the amount of masking that would extend above the arbitrarily chosen level of ambient noise (in spectrum level). The ambient chosen was offshore the island of North Bimini, Bahamas, where all the species shown are abundant. Audiograms selected from various authors (cited in text). See Fig. 3 and the text for further information.

function appears to be independent of environmental noise
and its is not unreasonable to suggest that a similar
physiological mechanism(s) is controlling the function in
most species.   The corresponding slopes below the regions
of peak sensitivity in these same fishes appear, on the
other hand,   to vary considerably at times and there
are strong suggestion that this function was dependent upon
the magnitude of ambient noise in that spectrum at the time
of testing. When noise is maintained at some constant level
at a given frequency, audiograms from different species
having similar regions of peak sensitivities show remarkably
similar slopes and associated sensitivities (Chapman, 1973;
Myrberg and Spires, 1977).

A contrasting study in this regard has been conducted,
however, by Tavolga and Wodinsky (1963) on nine species of
fishes common in the waters off Bimini, Bahamas.   These
authors reported an extraordinary 60 dB variation in sensitivity within the lowest frequency ranges tested (100 to 200
Hz , and yet many, if not all of the species examined were
from in the same general habitat. Large differentials were
even present among those species possessing similar peak
sensitivities at the same or similar frequencies. It is
possible that the fishes examined in that study were responding to the particle displacement component of the
acoustical stimulus at certain times and to the pressure
component at other times. Shifting of the modality used to
determine threshold sensitivity would most likely occur at
those low frequencies and this could explain at least some
of the variability noted. A somewhat more parsimonius explanation for such variation may be that differences in ambient noise were present in test tanks or possibly even
holding tanks (i.e., resulting in temporary threshold shifts,
Ha, 1968).

Closely related species appear to show similar hearing
abilities, so long as they are residing in a similar habitat
(e.g., the cod-like fishes of the family, Gadidae - Chapman,
1973; damselfishes of the genus Eupomacentrus - Myrberg and
Spires, 1977). Apparent exceptions are those squirrelfishes
of the genus Holocentrus, that were examined by Tavolga and
Wodinsky (1963).

Another trend noted among the audiograms of fishes is
one suggesting that the less sensitive a species is to sound
(i.e., the more distant its sensitivity is from probable influences of ambient noise), the steeper is its low frequency
slope. Although there, again, appear to be exceptions, this

can be noted in the figures supplied in various reviews, especially that by Popper and Fay (1973). This, in turn, suggests that the low frequency slopes of those species possessing good hearing abilities are being affected by some factor that is causing a "flattened" range of sensitivity in that portion of their audible spectrum. The thesis developed below is that ambient noise is that factor.

The long-standing question of sound localization in the acoustic farfield by fishes appears recently to have been answered, at least in part, for some species of teleosts and elasmobranch fishes. Strong evidence, from a number of recent studies, has favored such localization (Chapman, 1973; Enger et al., 1973; Myrberg et al., 1969, 1972, 1976; Nelson and Johnson, 1972; Nelson et al., 1969; Olsen, 1969; Sand, 1974; Schuijf, 1974; Schuijf and Siemelink, 1974; Schuijf et al., 1972). Present evidence also indicates that for fishes with gas bladders (e.g., the cod), sound-detection thresholds are below those required for sound-localization (Chapman, 1973). This relationship apparently differs, however, in fishes without gas bladders (e.g., the lemon shark), where only one level serves both functions (Banner, 1972).

ODONTOCETE CETACEANS AND PINNIPEDS

Relatively little quantitative information is available on the auditory sensitivity of marine mammals. Sound-detection thresholds have been obtained over a wide range of frequencies for only five species of odontocetes: four marine species--the bottlenose dolphin, Tursiops truncatus (Johnson, 1966), the harbor porpoise, Phocoena phocoena (Andersen, 1970), the common porpoise, Delphinus delphis (Belkovich and Solntseva, 1970; in Hall and Johnson, 1971) and the killer whale, Orcinus orca (Hall and Johnson, 1971), and one fresh-water species--the Amazon River dolphin, Inia geoffrensis (Jacobs and Hall, 1972). In each case only single animals were studied. Morozov has, however, recently confirmed Johnson's results on the bottlenose dolphin (W. Evans, pers. comm.). Five species of pinnipeds have been studied in like-manner. Schusterman (1974) and Schusterman et al. (1972) have established air and underwater audiograms for the single otariid studied, the California sea lion, Zalophus californianus, while Terhune and Ronald (1971 and 1972) and Møhl (1968) have carried out similar studies on two phocids, the harp seal, Pagophilus groenlandicus, and the harbor seal, Phoca vitulina, respectively. Additionally, underwater audiograms have been determined for the ringed seal, Pusa hispida (Terhune and Ronald, 1975a), and the gray

seal, Halichoerus grypus (Ridgway and Joyce, 1975); but as
in the case of the toothed whales, only one or, at most,
two individuals of each pinniped species were ever examined.
Nonetheless,the results, to date, show a number of clear correlates within the respective taxonomic groupings, as well
as among the group as a whole, despite differences in methodologies, available instrumentation, etc.

Although the various odontocetes have shown different
low frequency limits, all have demonstrated extremely high
sensitivity to a wide range of high frequency sounds. This
is exemplified by the harbor porpoise whose range extends
below 8 KHz and up to 140-150 KHz. All species appear to
have an extremely rapid cut-off in sensitivity at their upper
hearing limit (Andersen, 1970; Møhl, 1968).

A major portion of the regions of peak sensitivity fit
well the animals' own signal characteristics (e.g., Diercks
et al., 1973, but see Møhl and Andersen, 1973 and Watkins,
1974), this fact correlating well with that noted in those
species of fishes and other animal groups where similar data
are available. It is well known that marine mammals possess
a broad repertoire of widely differing sounds; but the most
"celebrated" ones, at least in the case of the odontocetes,
are the echolocation-clicks. There is little doubt that the
high frequency character of these brief, high energy pulses
is reflected in the high frequency sensitivity enjoyed by
these particular animals.

When considering pinniped audibility, the remarkable
similarity shown by the four phocid species (i.e., no difference in sensitivity greater than 20 dB at any frequency)
suggests that one may refer to a general "phocid"-audiogram.
Although Schusterman et al. (1972) considered that differences in the sensitivities existing between their
"otariid" curve and the only "phocid" curve available to
them at the time might actually be due to methodological differences and individual variation, some investigators presently believe otherwise(e.g., Terhune and Ronald, 1975a).
The underwater audiogram of the sea lion shows an extremely
broad region of highest sensitivity ($\sim$ - 15 dB /u.bar) between 1 and about 28 KHz. Between the latter point and
36 KHz, decreasing sensitivity reaches a steep slope of about
60 dB/octave. This same rate of loss in sensitivity was
also noted in the phocid seals between 45-50 and 64 KHz.
An interesting "leveling-off" of this steep slope, to about
12-14 dB/octave, occurred, however, in both groups near
their high frequency limit of hearing (the sea lion, near
32 KHz; the phocids, around 64 KHz). This phenomenon has

been considered by Møhl (1968) and Schusterman et al. (1972) to be "pseudo-hearing" of ultrasonic sound through bone conduction ("pseudo-hearing" meaning that only intensity differences are perceived among given sounds).

Although the differences between phocid and otariid hearing abilities may well be real, there appears to be sufficient similarity to justify a general comparison of audibility between them and the odontocetes. The odontocetes appear to have a 20 to 30 dB superiority over the pinnipeds in their respective regions of greatest sensitivity and, except for the killer whale (high frequency cut-off similar to that of the sea lion, i.e., about 30 KHz), their audibility extends 1 to 2.5 octaves beyond that of the pinnipeds (ignoring here the region of pseudo-hearing; see Schusterman et al., 1972). These differences are, no doubt, due in large part to the echolocating capability of at least certain odontocetes and as Schusterman et al., have appropriately stated: "...It does not seem unreasonable to assume that echolocation was a major source of selective pressure for excellent hearing sensitivity at frequencies above 60 KHz..." The development of high frequency audibility in seals may, indeed, have been confined by their limited use of echolocation, as well as by their amphibious nature which requires that they possess relatively good hearing both underwater and in air (for review, see Schusterman, 1974).

AUDIBILITY AND ENVIRONMENT NOISE: THE RELATIONSHIPS

When considering the effects of ocean noise upon audibility among both the marine fishes and mammals, the frequency range involved is extensive. This is due to the fact that the most important region of sound detection in most fishes rests between about 40 and 1000 Hz, while that same region in the pinnipeds is located between 500 Hz and 30 to 45 KHz, and in the odontocetes, between 8 KHz and 120 to 145 KHz.

I have, therefore, taken the liberty of dividing this portion of the report into three sections. The first deals with those fishes whose hearing sensitivity rests in what may be termed the extremely low register, i.e., between 10 and 500 Hz. These fishes, including the cod and its relatives (e.g., the haddock, pollock and ling), the toadfish and the sharks, all appear keenly adapted to this particular range of frequencies, the teleosts because of their own respective signal characteristics (Brawn, 1961; Fish and Offutt, 1972; Gray and Winn, 1961; Winn, 1967) and the elasmobranchs because of the various sounds produced by their prey (Myrberg et al., 1976; Myrberg, in press, Nelson and Gruber, 1963; Nelson and Johnson, 1976).

The second group is made up of those marine fishes whose peak sensitivities extend from about 200 to 1000 Hz. The majority of teleost fishes appears to belong to this group and many of these, in turn, belong also to the advanced and highly radiated taxonomic order, the Perciformes. Although there certainly are glaring exceptions, this rather genera-lized group constitutes the most speciose and populous animals of the shallow water. Their domain includes rocky coast-lines, regions of mangroves and seagrasses and the coral reef.

Based on our present knowledge, the third group – the marine mammals – can be divided into three sub-groups: the otariid and phocid pinnipeds and the odontocetes (the delp-hinids and the single phocoenid studied appear to possess more similarities than differences in their auditory capabilities). All three sub-groups have audio-sensitivities that extend be-yond 20 KHz. Their habitats include generally the coastal and littoral regions, although in the case of two delphinids – the common porpoise and the killer whale, the preferred habi-tats appear to be over deep oceanic waters (former) and either over deep water or the littoral region (latter) (Evans, 1973).

GROUP 1

Figure 3 depicts various ambient noise effects in re-lation to auditory thresholds from four selected species whose sensitivities span most of the frequency spectrum shown. It is clear that the threshold sensitivities obtained by Chapman (1973) for the cod, the haddock, and the pollock (all shown in the figure) are close to, and could easily be ex-ceeded by the ambient noise. These three thresholds (a fourth, not shown, was from the ling-cod), are especially informative since they were all obtained directly in the field off the Scottish coast and their associated ambient levels fitted well those recorded by Piggott (1964), also off the Scottish coast. Chapman mentioned that there was a direct correlation between thresholds for the haddock and the spectrum level of the noise. This same correlation was also reported for the cod (Chapman and Hawkins, 1973); and based on many determinations, sea state appeared to be the causa-tive factor. Their results showed unquestionably that masking of threshold by ambient noise, although negligible in calm sea conditions (i.e., sea state 0), invariably occurred

at higher sea states. The frequencies not affected by such
noise levels were invariably those at, or near, the respective
ends of the hearing curves where the animals' sensitivity was
sufficiently low to preclude further impairment, or masking,
by noise. The remaining threshold curve, depicted in Fig. 3,
is from the toadfish. It was initially determined by Fish and
Offutt (1972) in the laboratory with specific points being
later confirmed in the field. The species has low sensitivity
compared to that of the gadoid-fishes at similar frequencies;
but it is likely that reasonable levels of ambient noise would
also mask its most sensitive region.

In the above studies, the investigators also found that
a specific relationship apparently existed between a given
frequency-threshold and the spectrum level noise at that same
point. This was subsequently investigated in the cod by
Hawkins and Chapman (1975), following the leads by Tavolga
(1967) and Buerkle (1968, 1969) that ambient noise could,
indeed, mask the detection of sounds by fishes.

An equally illuminating study by Banner(1972) has also
demonstrated that the hearing capability of lemon sharks,
Negaprion brevirostris, depends on a remarkably consistent relationship between signal and ambient noise. Signal to
noise (spectrum level) capability measured in the laboratory
at two different frequencies was shown to be statistically
identical to the signal to noise response levels shown by
free-ranging lemon sharks in the field. Since he had previously shown (Banner, 1967) that hearing thresholds for the
species were dependent on particle motion rather than pressure, it is interesting that identical signal to noise capabilities were found, despite the fact that the signals and
associated noise in the field apparently were directed
vertically (shallow water) whereas those in the laboratory
were predominantly horizontal.

The   above studies, plus others by Fay (1974), Ha
(1968) and Tavolga (1974b) have established beyond question
that, in fishes, a specific relationship does, indeed, exist
between the spectrum level of noise at a given frequency and
the detectibility of a signal at, or near, that frequency.
This relationship does not hold, of course, when detection is
not impaired by noise, e.g., when the level of background
noise is decreased after absolute threshold is reached.

The relationship between a masked threshold determina-
tion and its corresponding spectrum level noise has been
variously termed: signal/noise ratio (Tavolga, 1974b),

threshold: noise ratio (Hawkins and Chapman, 1975); and the critical ratio - CR (Fay, 1974; Ha, 1968; Johnson, 1968; Terhune and Ronald, 1975b; and Albers; in Ha, 1968). Although CRs' (in Hz) were interpreted by Hawkins and Chapman (1975) to be calculated values for the critical band rather than empirically derived values, it is apparent that others use the term to refer specifically to the relationship or ratio (in dB) described above.

Perusual of the literature regarding the constancy of the above ratios resulted in the summary provided in Table 1. These specific threshold: spectrum level noise values, or critical ratios (CRs') ranged from 14 to 29 for all fishes, except for the goldfish (the single "specialist" listed). A clearly suggestive trend is also apparent, regarding the values for a given species, i.e., a general (but not universal) increase as frequency increases, e.g., cod, (Buerkle, 1968). This same directional trend is also seen among the values for man and the marine mammals. Finally, as Fay (1974) has pointed out, since the CRs' are indirect estimates of the frequency range over which two stimuli interfere with one another (Licklider, 1959), a growth in the size of the CR with frequency indicates that the resolving power of the system is declining.

Based on these data, specific values were assigned various frequencies across the "best hearing" spectrum of the cod. These were then added to the spectrum level noise of an arbitrary sea state (slightly less than sea state 2) at the same frequencies. The resulting simulation is shown in Figure 3, after adding to the lower frequencies a light level of traffic (industrial) noise that coincided with the selected sea state at the frequency of greatest sensitivity. From this simulation, it was apparent that if the thresholds had been determined under the simulated conditions, a line of masked thresholds would have resulted (=line drawn below CR=17-24). It is noteworthy that the unique shape of the simulated hearing curve follows almost precisely the curves of the actual thresholds obtained for the cod and pollock by Chapman (1973). This suggests that thresholds in this region may well be affected not only by wind, but also by traffic (or industrial) noise, as well. The particular simulation used is a reasonable one from the standpoint of ambient noise and it clearly substantiates the statement made by Chapman and Hawkins (1973) that only a calm sea produced reliable thresholds. Even at sea state 1, with extremely light traffic noise, the cod's threshold would probably have been masked. It is also clear that any increase in ambient noise

at such low frequencies should have a greater effect on
audibility than a similar increase at higher frequencies
(because of the general slope). This probably accounts for
the wider and flatter region of relatively good sensitivity
in this portion of the spectrum compared to that seen in
higher regions. The above statements (and those which follow)
imply, of course, that fishes do experience hearing loss.
This particular point, recently examined in a few species,
will be briefly discussed later in this report (p. 28, ms.).

Before leaving this section, it should be pointed out
that in those animals sensitive to particle motion (e.g.,
sharks), directionality of ambient noise may also be an important factor. A fish with directionally sensitive hearing
might actually discriminate a source from background noise
by taking advantage of differences in the orientation of
source and the ambient noise particle motion. If so, one
must account, however, for near-field effects of both types
of sound (see Banner, 1968) and under such circumstances, a
displacement-sensitive receptor might be at a disadvantage
unless the direction of the particle motion was not the same
as that of the source. In any case, it is apparent that both
signal and noise must be measured in the same manner and under
the same conditions as experienced by the test animal.

GROUP 2

Audiograms of ten species of marine fishes (Myrberg
and Spires, 1977; Tavolga, 1974a; Tavolga and Wodinsky, 1963)
are shown in Figure 4, along with ambient noise effects within the selected spectrum. The spectrum level noise of a relatively calm sea state (approaching 1 - over a depth of 25 m;
Myrberg et al., 1969) at Bimini, Bahamas is also added since
all of the included species are found in the waters off that
island. It is apparent that wind (sea state) and traffic
noise are again the major determinants of sea noise in the
spectrum being considered. Biological sources of noise
would add, of course, to any effect or result noted below.

Those species having their greatest sensitivity in the
region designated for this group have shown, with only one
exception to date, that their critical ratios fall between
19 and 26 dB (Table 1 - following page). These values are
somewhat larger than those obtained from the few species that
have been examined in Group 1. Although the groupings are
based largely upon convenience and many more values are
needed to confirm the indicative trend, it appears that those
species of Group 2 possess a somewhat reduced resolving power

TABLE 1

Auditory threshold: spectrum-level noise (or critical ratio) (dB) in the presence of masking noise.

H_z	4×10^1			1×10^2							1×10^3					1.6×10^4		
	5	6	7	1.5	2	3	4	5	7	8	1.2	2	4	6	8	3.2	6	8
Man in air (Hawkins and Stevens '50)				19	18	16			17		18	20	23		27			
Marine mammals																		
Ringed seal (Terhune and Ronald '75b)													30		32	34	35	
Bottlenose dolphin (Johnson '68)														22	25	32	37	39
Marine fishes																		
Cod (Hawkins and Chapman '75)	17	16		16	19	21	21											
Cod (Buerkle '68)	18	24																
Ling (Chapman '73)	17		21	24	25	25	26											
Haddock (Chapman '73)	21	25		21	23	22												
Pollack (Chapman '73)	21			22		26												
Blue-striped grunt (Tavolga '74b)								24										
Pinfish (Tavolga '74b)						20		29										
Longspine squirrelfish (Tavolga '74b)								14		25								
Lemon shark (Banner '72)	21																	
Freshwater fishes																		
Goldfish (Tavolga '74b)	8			14	22	23	23	23	23									
Goldfish (Fay '74)				13	17	19	23	23	22		25							
Black-chinned mouthbrooder (Tavolga '74b)						23		24										

185

compared to those species having their peak sensitivities at lower frequencies. Based on the reasoning by Fletcher (1940) and subsequent authors dealing with mammalian hearing, the maximum band of frequencies which can effectively mask a given tone, i.e., the critical band, has been recently studied in fishes by various investigators. Although controversy and speculation presently abound regarding the usefulness of the concept in fishes, I agree with Hawkins and Chapman (1975) that there are advantages to be gained by describing the masking function in terms of an effective band of finite width.

Since the level of noise within a band width Δf_c is equal to the spectrum level plus 10 log $_{10}\Delta f$, then Δf_c equals the critical band. By using the CRs' provided in Table 1, it can be easily determined that such bands, if present, would generally be 40 to 160 Hz wide (depending on the frequency in question) for frequencies below 100 Hz and from 80 to 400 Hz wide (depending on frequency) for frequencies between 100 and 1000 Hz.

When a simulation, such as that carried out in Fig. 3, is applied to the data provided in Figure 4, using as its base the noise level for a relatively calm sea state off Bimini, it is again readily apparent that the hearing abilities of the most sensitive species will be impaired to a level quite similar to those initially far less sensitive than they. Also, any increase in ambient noise above that of a calm sea probably affects all the thresholds depicted in Figure 4. The simulation points out that statements concerning auditory capacities of fishes must be related to background noise. Only through a knowledge of the relationship that exists between a given threshold and its corresponding spectrum level noise can one make meaningful statements concerning the auditory capabilities of fishes in their natural environment.

A comparison of the respective shape of the audiograms shown in Figure 4 shows that the more sensitive a species is to sound, the less steep is its ascending slope of sensitivity in the lower frequencies. This indicates that even under quiet conditions of the laboratory, the thresholds in that region, for the most sensitive species, are still being influenced by extraneous noise.

GROUP 3

Evidence continues to accumulate showing that all

marine mammals produce a variety of sounds. Hearing capability has been examined, however, among only certain odontocetes and pinnipeds. Representative audiograms are shown in Figure 5 and their divergence, one from the other, reflects the taxonomic divergence of the groups involved. Although these mammals certainly perceive underwater signals below 500 Hz, as well as a variety of air-borne sounds, discussion will be limited to their higher-frequency, underwater capability.

The audiogram of the California sea lion is the one available for the family Otariidae (Schusterman et al., 1972). Its essentially flattened nature between 1 and 30 KHz indicates that sensitivity throughout that region was not affected by noise during testing. This stands in contrast to the available audiograms on the harbor seal, representing the Phocidae and that of the bottlenose dolphin, representing the odontocetes (Johnson, 1966). It is probably not coincidental that the slope of decreasing sensitivity for the harbor seal parallels reasonably well the slope of ambient noise in the same region. Although the equivalent slope for the dolphin departs somewhat from that normally associated with noise, it, nevertheless, indicates that some type of traffic (or industrial) noise may well have affected sensitivity during testing.

The 15 dB increase in peak sensitivity of the phocid threshold over that of the otariid suggests a real difference in capability between members of these two families. This is borne out especially well by the great similarity in sensitivity, recorded from all the phocids studied thus far. When considering the effects of ambient noise on hearing in these two groups, taxonomic or physiological reality, however, may well be surrogated by the reality of the underwater world in which these animals live. This may well be the case here since a recent communication by Schusterman (pers. comm.) points out that data being obtained on the northern fur seal, Callorhinus ursinus suggests no difference in absolute sensitivity from Phoca.

Critical ratios have been determined for only one pinniped; Terhune and Ronald's elegant study (1975b) of the ringed seal provides the unique opportunity to do a bit more than speculate about the possible effects of noise upon the hearing abilities of these animals and their near relatives. Results showed that the CRs' varied from 30 to 35 dB between 4 and 32 KHz (Table 1). When these values (increasing values with higher frequencies) are added to a relatively

low ambient level such as that provided by a sea state 1,
the actual thresholds for the ringed seal, as well as those
of the harbor seal (shown in Fig. 5) and the remaining phocids become masked - with the resulting thresholds approaching that of the less sensitive sea lion. This can be seen
in the simulation provided in Figure 5 (a similar interpretation is made also by Terhune and Ronald). Slightly
different CRs' were used so that the simulation might also
be applicable to the odontocetes.

If masking takes place under a calm sea, the effects
of any traffic or industrial noise above that level should
add to a further decrease in sensitivity. Those sources
plus that of sea state are reduced to extremely low levels,
however, when frequencies beyond 100 KHz are reached. Rain
noise, though an intermittent variable, may, however, "play
havoc" with hearing during certain times of the day or
during certain seasons of the year. When critical ratios
are taken into account during periods of even moderate rain,
hearing in phocids could decrease to levels of +5 to +10 dB/
ubar over a broad range of frequencies. And since the CRs'
of the odontocetes are not greatly dissimilar from those of
the phocids, it is reasonable to asume that those of the
otariids will also be in the same "ball-park". This latter
group would most probably also be affected, but less so.

Sound emissions by the otariid and phocid pinnipeds
have been described by various authors (Evans, 1967; Poulter
1968; Schevill et al.,1963; Schusterman,1968; Schusterman
and Balliet, 1969; Schusterman et al., 1970). Many of the
sounds emitted in air are also emitted underwater. Although there is much energy below 2 KHz, energy is also
present in the 8 to 12 KHz range (e.g., barks - Otariidae;
roars, moos, yodels, hums and clicks - Phocidae). Within
that range,ambient noise will probably regulate sensitivity.
This particular point will be more thoroughly discussed
later.

The high sensitivity of various odontocetes at frequencies between 50 and 145 KHz correlates well with the
known echolocating emissions by these animals (see Evans,
1973). Although rain noise (Fig. 5) and certain biological
sounds (e.g., shrimp, see Fig. 1) do constitute major components of the noise at frequencies between 20 and 100 KHz
(and even higher, see Au et al., 1974), the major sustaining
factor affecting sensitivity in these animals should be
their critical ratios. Fortunately, we again have a
single, but important study that deals specifically with
these values in the bottlenose dolphin (Johnson, 1968). The

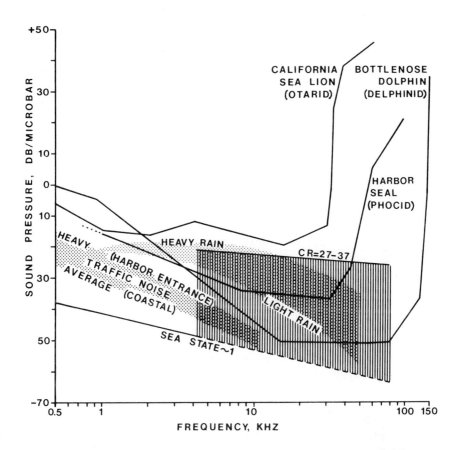

Figure 5 - High frequency ambient noise and its probable masking effect on the hearing abilities of selected marine mammals whose peak sensitivities are found within that spectrum. Audiograms were redrawn from the works of several authors (cited in text). The hatched area is the region chosen to show the amount of masking that would extend above the arbitrarily chosen level of ambient noise (in spectrum level) for the critical ratios (CR), given.
See Figs. 2, 3 and text for additional information.

CRs' reported in that study are provided also in Table 1 and
they were used, in part, for the simulation provided in
Figure 5. Based on the very high values of these CRs', it
is apparent that the actual thresholds of these animals will
also be masked, with resulting sensitivity being only
slightly below those of the pinnipeds. Although the degree
of masking is considerable, the reduced sensitivity appears
still fully sufficient for the task of echolocating (see
below). At frequencies below 15 KHz, background noise increases, but this is compensated by decreasing values of the
CR. This interesting "interplay" between an animals's sensitivity and specific factors in its environment may well
have implications relative to the numerous sounds, other
than those of echolocation, produced by these highly soniferous animals.

AUDIBILITY AND ENVIRONMENTAL NOISE: THE IMPLICATIONS

A. Marine fishes

It is not unreasonable to assume that in the shallow,
coastal regions of the oceans, sound reception by the
majority of fishes is limited by environmental noise. Free-
dom from such an influence is probably "enjoyed" by members
of a given species only at the extremes of their hearing
range, where physiological restraints then set the limits.

Although various studies in the past referred to a
rather specific relationship existing between threshold
sensitivities and spectrum level noise - i.e., a 20 dB
(+ a few dB) difference at frequencies of a few hundred
hertz (Banner, 1967; Buerkle, 1969; Cahn et al., 1969;
Myrberg et al., 1969; Nelson, 1967), consideration of ambient
noise and particularly its spectrum levels was, until recently, often relegated to a non-empirical factor that had
to be only "discussed away". Now that this specific relationship has been firmly established by a number of elegant studies, it is clear that this "critical ratio" is
a clear indicator of masking. If its value at any given frequency remains relatively constant, sensitivity is most probably being masked; if its value exceeds the point of constancy, masking is unlikely. The reason for the older
studies often referring to a rather similar value is now
also obvious - they (and others like them) were dealing with
masked sensitivities.

While discussing masking by ambient noise in fishes,
Tavolga (1967) noted that its effects in nature were probably minor, except under highly unusual conditions for a

few species with unusually good hearing. I plead here the
contrary case: the effects of ambient noise are paramount
to hearing abilities except under the often unusual condition
of a very calm sea with less than a "whisper" of mechanical
(i.e., traffic or industrial) noise. When seas of state 1
or higher are noted, when traffic or industrial noise is
apparent, or when other low frequency sound sources such
as shifting bottoms, tidal currents, or noisy animals are
encountered, sound reception in all, but the "deafest",
fishes is probably impaired - the degree of such impairment
being directly dependent upon the noise-level attained.
This particular point has rather far ranging implications.

Underwater ambient noise is, of course, characterized
by a Gaussian amplitude spectrum, while sounds of most, if
not all, marine vertebrates often possess recognizable frequency spectra, along with appropriate modulations. Therefore, it is likely that most species which rely on acoustic
reception can utilize some form of signal processing so as
to perceive specific (biologically important) signals even
at low amplitudes. The spectral content of given signals,
their redundancy and relative stability all provide ways
and means to accomplish such feats. Nevertheless, in the
following section, I shall speculate upon findings from
pure tone sensitivities with inferences made about actual
sounds and their reception by the animals concerned.
Certain readers will probably consider such action unwar-
ranted, because of the absence of confirming evidence. This
point is especially justified in the case of the marine
mammals since their hearing abilities cover such a broad
frequency range that differential masking may actually pose
little or no problem to them. Yet, at this stage in our
development and with our present knowledge, I don't think
it is unreasonable to suggest that the present possibility
that ambient noise can affect biological signals appears
to be changing to probabilities, the levels of which must be
determined in the future by concerned scientists. By using
here an extremely simplified and a conservative approach,
the figures provided below would, of course, represent ac-
tuality only if signal processing was at an absolute minimum.
If, on the other hand, complexity characterizes such processing, sound-detection limits would clearly exceed those
given below. For the purposes of this symposium, I'm going
"far out on a limb of speculation" and the limb will most
assuredly be "cut off" behind me; how close the cut to the
trunk (or to my position) remains to be seen.

Although we know that the vast majority of fishes
make sounds, detailed knowledge about the biological significance of sound production is scant, except in a few in-

stances. In Table 2 (following page), four such "instances"
are listed, along with certain facts and some rather conservative speculations about sound production and reception.
Simple calculations have, accordingly, revealed additional
information relative not only to "acoustical biology" but
also to other problems seemingly unrelated to the modality
of sound.

The major bio-acoustical function in the case of the
first three species, listed in Table 2, appears to be intraspecific communication. Sounds "inform" conspecifics that
the sender is physiologically "ready" for specific types of
activities, be they reproductive, agonistic or what-have-
you. Among nearby conspecifics, there will be receivers
who, in turn, are also physiologically "ready" for the same
types of activities and it is they who respond. Since it
is imperative in such situations that the "message" from
the sender reach a receiver, the actual distance between
such individuals is important. If ambient noise adversely
affects the level of the signal (message), communication
is precluded or, at least, interrupted; and when, as in the
case of the first example listed in Table 2, the bicolor
damselfish, such a signal is involved in initiating and
maintaining the courtship ritual in the species (see
Myrberg, 1972; Myrberg and Spires, 1972) reproductive activity may well be curtailed.

The fourth example, a shark, being a predator would
find it to his definite advantage if he could hear and interpret the sounds of his prey. A number of studies have
in fact, shown that sharks, as a group, clearly use their
sense of hearing for that purpose (e.g., Banner, 1972;
Myrberg, in press; Myrberg et al., 1972, 1976; Nelson and
Gruber, 1963; Nelson and Johnson, 1972, 1976). Again, such
information transfer, being distance-related, can be affected by the level of ambient noise. If prey detection
and localization is dependent on the hearing modality, the
"stake" has now become the severest of games - survival.

All values within the second through the fourth
columns of Table 2 are published except where asterisks
are used. Although an audiogram is available for the bicolor damselfish (Ha, 1973), a study dealing with proven
masked thresholds has not been done. Therefore, an estimated CR was supplied, its value being the mean of those
which have been obtained on other teleosts at 500 Hz. This
same procedure was also used to supply an appropriate CR to
the toadfish for the same reason. There is no published
sound-source level for the longspine squirrelfish and so
an arbitrary level was also assigned in that instance, it

TABLE II

Estimated sound-detection-distances under different ocean-noise conditions for selected species of marine fishes. Conspecific source levels used in all calculations, except for those involving the lemon shark (see text); audio frequencies selected from regions of peak energy for the respective sound sources.

Species	Sound-source level (dB/µbar re 1 m)	Selected audio-frequency (Hz)	Audio threshold; spectrum level noise ratio (dB)	At Sea State	Most sensitive audio threshold (dB/µbar)	Estimated maximum detection distance (meters)	At traffic level (sea state 1)	Most sensitive audio threshold (dB/µbar)	Estimated maximum detection distance (meters)
Eupomacentrus partitus Bicolor damselfish	+ 7	500	23*	1	-12	9	Light	-12	9
				2	- 7	5	Average	- 3	3.5
							Heavy	+ 6	1
Holocentrus rufus Longspine squirrelfish	+13*	600	23	1	-16	30	Light	-13	20
				2	- 8	12	Average	- 4	7.5
							Heavy	+ 5	1.5
Opsanus tau Toadfish	+35	100	17*	1	- 2	75	Light	- 2	75
				2	- 2	75	Average	- 2	75
							Heavy	+ 7	30
Negaprion brevirostris Lemon shark	+30*	300	20	1	-13	150	Light	-12	130
				2	-10	105	Average	- 3	50
							Heavy	+ 6	20

*assumed values (see text)

being twice the level of the sound produced by the bicolor
damselfish (personal experience with both sounds in the
field indicates that the level is reasonable). The sound-
source level used for the lemon shark is that of a natural
prey-sound measured and used by Nelson and Johnson (1970) in
attraction studies on sharks ("stampeding" bonefish, Albula
vulpes).

The calculated values in the tables are based on established audiograms for each species; and by applying the
appropriate CRs' to levels of the varying types of noise,
maximum-detection distances (in meters)have been determined.
Such determinations are the first attempts to provide a most
conservative idea of how far a sender can "send" its message
so as to reach an appropriate receiver. Since fishes are
apparently unable to vary appreciably the sound-level of
their signals, the distance over which a sender transmits
its signal to an appropriate receiver should remain relatively constant (i.e., all else being equal).

In the case of the bicolor damselfish, large, repro-
ductively-active males within a colony maintain territories
of such a size that the nearest large males are between 3
and 5 m away. Since the courtship sounds of one such male
in its territory causes nearby males to begin competitive
courtship rituals, it's noteworthy that such distances also
happen to be the maximum distances of detection for their
respective sounds under conditions that are normal for our
regions(i.e., sea state 2 with a moderate level of traffic).

The distance over which the sounds of the longspine
squirrelfish travel exceeds that normally defended by such
animals. It's reasonable to assume, however, that their
sounds are used for functions that demand either operative
distances of 8 to 12 m (for conditions normal to our region)
or high levels close to their residences. Toadfish produce
some of the loudest sounds made by fishes. Since males often reside near one another, indications are that the corresponding levels are "used" for the purpose of attracting
females from over a wide area. Except under conditions of
heavy traffic, that area should have a radius of approximately 75 m under reasonably quiet sea states.

In the case of attracting a shark to the source of a
relatively loud, biologically-interesting sound, both detection and localization (see Banner, 1972) should occur
between 100 and 150 m away under reasonable sea states and
seas, the maximum proven distance over which other carchar-
hinid sharks have been attracted to similar sounds is be-

tween 300 and 400 m. Such distances fit well the values
assigned to slightly higher seas.

Thus, based only on these four examples, social function and even predation may well be dependent for their
success on the level of environmental noise. Signals are
enhanced as noise is reduced and vice versa; slight changes
in environmental states result in clear changes in receptive
fields.

Although sound reception has been emphasized in this
section some evidence has accumulated that sound transmission, itself, is adversely affected by noise. Winn
(1967) obtained, for example, clear evidence during a
field study of vocal facilitation in the toadfish, that
recorded members of the species significantly reduced their
calling rate when background noise, normal to the area, was
played back at levels far less than that of the species'
own sounds (which increased calling rates). This finding
indicates that fishes can perceive differences in sea noise.
The fact that most fishes are limited in their hearing only
by the level of ambient noise also suggests that they are
able to detect differences in that noise.

Still another implication of the results, summarized
here, has to do with commercial fisheries. The sounds of
fishing vessels and their associated gear can probably be
heard by their potential catch when levels (now "signals")
reach some point greater than the background noise (see
Maniwa, 1971; Olsen, 1971). Since response thresholds and
those of detection need not necessarily be the same, further
investigations regarding this important factor in coastal
and high-seas fisheries is clearly warranted.

Similarly, important questions to be answered in the
future are whether marine animals suffer temporary or even
permanent hearing loss when subjected to unusually noisy
environments for long periods of time or do they avoid or
leave such localities. Although almost no information is
presently available regarding these questions, Popper and
Clarke (1976) have recently shown that goldfish experience
temporary hearing loss (about 24 hrs.) after being exposed
to intense signals of approximately +49 dB/ubar (equivalent
to +123 dB/0.0002 ubar) for four hours. Such losses are
manifested by appropriate temporary threshold shifts.
Similar findings, using less intense signals, have also been
shown for the lane snapper, Lutjanus synagris by Ha (1968).

The final implication to be mentioned is that based on Banner and Hyatt's study (1973) of acoustical noise and its effects on the development of two common estuarine fishes (see Figure 6). These authors noted, under controlled testing, that viability of the edges of Cyprinodon variegatus was significantly reduced in aquaria when a low frequency (40 to 1000 Hz) noise level, approximately 40 to 50 dB above that normally encountered in the natural habitat, was maintained over a number of consecutive days. Such lethal effects were apparently restricted to embryonic stages since fry exposed to such levels experienced no losses. Growth rates in that species, as well as in Fundulus similis, were, however, significantly less than those noted when noise levels were maintained about 20 dB less during the same time period. These results are reminiscent of those pathological effects of excessive noise that have been observed in various mammals – including man – during recent years (e.g., Athey, 1970; Fry et al., 1970; Miller et al., 1971). This problem area will unquestionably expand in importance as economic development continues to increase within, or adjacent to, aquatic areas.

THE MARINE MAMMALS

The comments made on page [16] relative to the theoretical limits for sound detection, are especially appropriate in the context of marine mammal activities. The simulations provided below are meant only to emphasize concern for the problems of sound detection by these animals and to touch on the complexity of the total problem that might face these animals in the acoustical domain.

Table 3 and 4 follow basically the same format as that used in Table 2. To standardize sound source levels, all values have been expressed in peak to peak values (to obtain approximate rms values, subtract 20 $[\log_{10} 2(1/\sqrt{2})]$ from the assigned values).

The audio-frequency selected for the harbor seal (Table 3) falls within the region of peak energy for its buzz-like call. Those used for the bottlenose dolphin (Table 4) are, like-wise, within the regions of peak energy for the respective echolocating clicks at different sound-source levels. Two-way transmission was considered in all calculations regarding the dolphin. An arbitrary CR value was assigned to the harbor seal, it being slightly below that of its relative, the ringed seal. The absorption coefficient, used in the calculations of Table 4, was 0.05 dB/m (Urick, 1975); Au et al., (1974) used a similar value which agreed with theoretical considerations.

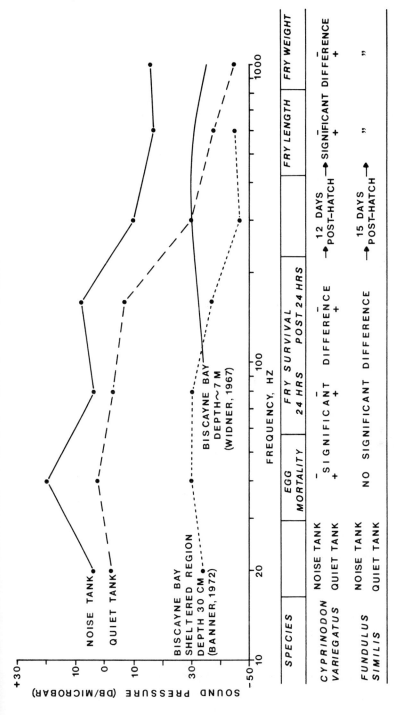

Figure 6 – The effects of environmental noise on marine animals.

197

TABLE III

Estimated echo-location-limits, in meters, under different ocean-noise conditions for the bottlenose dolphin, Tursiops truncatus. Capability of varying sound-source level assumed; audio-frequencies selected from regions of peak energy for the respective sound-source levels (see text); absorption coefficient ~0.05 dB/1 m (m.d.d. = maximum detection distance).

Source level (average p-p) (dB/μbar re 1m)	Selected audio-frequency (KHz)	Audio-threshold: spectrum level noise ratio (dB)	At sea state	Audio threshold (dB/μbar)	Est. m.d.d. (m.)	At traffic level (sea state 1)	Audio threshold (dB/μbar)	Est. m.d.d. (m.)	At rain level (sea state 1)	Audio threshold (dB/μbar)	Est. m.d.d. (m.)
+ 45	30	32	1	-27	60	Average	-25	55	Light	-17	40
			2	-23	55	Heavy	-22	45	Heavy	+ 3	15
			.... Sustained noise by snapping shrimp	+ 2	10						
+ 85	50	37	1	-26	260	N.A.	...	...	Light	-10	160
			2	-22	215				Heavy	+ 1	90
			.... Sustained noise by snapping shrimp	- 3	105						
+120	110	39	1	-29	640	N.A.	...	...	Heavy	-15	520
			2	-25	600						
			.... Sustained noise by snapping shrimp	- 1	340						

TABLE IV

Estimated sound-detection-distances under different ocean-noise conditions for the common seal, _Phoca vitulina_. Selected audio-frequency, 9 KHz (see text); audio threshold: spectrum level noise ratio = 30dB.

Sound-source level (p-p) in dB/μbar re 1m . . . . +38
(e.g., conspecific)

	At sea-state	
	1	2
Most sensitive threshold (dB/μbar)	-22	-15
Estimated maximum detection distance (meters)	1000	500

	At sea-state 1			
	Traffic level		Rain level	
	Average	Heavy	Light	Heavy
Most sensitive threshold (dB/μbar)	-19	-14	-3	+9
Estimated maximum detection distance (meters)	750	425	120	30

199

The simulation, shown in Table 3, involves the detection of a signal by a conspecific at a distance under two rather calm sea states and under differing levels of traffic and rain noise. Low sea states were used to demonstrate capability, it being reduced considerably at higher states. Threshold values were based on the established audiogram of the species (Møhl, 1968).

It is apparent that signal detection by the harbor seal (and probably by its near relatives) may also be dependent on background noise. Maximum detection-distance for reception in a relatively calm sea (state 1) is estimated at 1000 m. Although traffic (or industrial) noise affects reception, it is small compared to the loss in sensitivity during periods of rain. When such conditions become severe, communication is probably reduced to a few 10s' of meters.

In the case of the bottlenose dolphin (Table 4), difficulties were initially encountered when attempting to produce a simple table. Results from the literature provided quite different sound-source levels for the species; and based on the expertise of the observers, choosing a single "correct" level was impossible. Rather, it was obvious that members of the species do not "hold" to a certain level of sound production, but actually vary the magnitude of their signals so as to conform to the situation that confronts them. It was also noted that a corresponding change in the level of the echolocation-clicks resulted in a change in the spectrum of peak energy. Based on the comments by Au et al., (1974), the effects of sustained noise by snapping shrimp were also considered (sound-pressure level is based on measurements in Kaneohe Bay).

It's clear that 40 dB increases in sound-source levels would increase effective echolocating distances by 3 to 4 times. This is a reasonable factor, based on spreading and absorption. At the quietest source level, detection distances should be reduced considerably, especially in the presence of high levels of biological noise. At the two highest levels used in the simulation, the effects of traffic (or industrial) noise upon sensitivity appear essentially nil.

The maximum echolocating distance under a relatively calm sea was estimated to be somewhere around 650 m. Under the same sea conditions, but in regions of high-level biological sources, such as snapping shrimp, that distance should drop almost one-half.

Thus, hearing by odontocetes may also be affected by
environmental noise; and if so, the effective spectrum is
quite different from that which affects hearing in marine
fishes.   It is also clear that extreme high frequencies are
little affected by environmental noise; but the animals
which are sensitive to that spectrum will probably have to
deal with their own "noise", such as the limits that are
imposed by underlying physiological mechanisms.

Before ending this final section, it should be mentioned that the great whales of the order, Mysticeti, produce a wide variety of sounds, ranging from the extremely
complex "songs" of the humpback (Winn and Perkins, 1970) to
the "20 cycle-pulse" of the blue whale (Cummings and Thompson, 1971).   Since the sound levels produced by these mam-
moth creatures are truly astounding, it is only reasonable
to assume that, at least, certain of the sounds are used
for purposes of communication.   If such is the case, their
extremely low frequency vocalizations may well be affected
by traffic noise, be they in coastal regions or on the
high seas.   Also, with increasing industrial and related
noise, it is probable that the same processes that affect
the tiniest of fishes may well affect the largest animals
on this planet.

ACKNOWLEDGEMENTS

I am extremely grateful to many individuals for their
insight and critical advice to this project.   These include Arnold Banner, Harry A. De Ferrari, William E. Evans,
Charles R. Gordon, Joseph D. Richard, Ronald J. Schusterman,
Juanita Y. Spires, William N. Tavolga, William F. Watkins
and Gordon M. Wenz.   I have incorporated most recommendations
of these and other scientists and all have strengthened the
final story.   Certain recommendations were not followed,
however; and responsibility for any factual errors or con-
troversial subjects that might appear in the report rests
solely with the author.   This work relates to Department
of the Navy Contract N00014-76-C-0142 issued by the Office
of Naval Research.   The United States Government has a
royalty-free license throughout the world in all copyright-
able material contained herein.

LITERATURE CITED

1. Albers, V.M. 1965. Underwater Acoustics Handbook II.
    Pennsylvania State University Press, University Park.
    356 pp.
2. Andersen, S. 1970. Auditory sensitivity of the har-
    bour porpoise, Phocoena phocoena. pp. 255-258. In:
    G. Pilleri (ed.), Investigations on Cetacea. Vol. 2,
    Hirnanat. Inst., Berne.
3. Arase, E.M. and T. Arase. 1966. Correlation of am-
    bient sea noise. J. Acoust. Soc. Amer. 40:205.
4. Athey, S.W. 1970. Acoustics technology - a survey.
    National Aeronautics and Space Administration,
    Washington, D.C.   139 pp.
5. Au, W.W.L., R.W. Floyd, R.H. Penner and A.E. Murchison
    1974. Measurement of echolocation signals of the
    Atlantic bottlenose dolphin, Tursiops truncatus
    Montagu, in open waters. J. Acoust. Soc. Amer. 56(4):
    1280-1290.
6. Banner, A. 1967. Evidence of sensitivity to acoustic
    displacements in the lemon shark, Negaprion breviro-
    stris (Poey). pp. 265-273. In: P. Cahn (ed.),
    Lateral Line Detectors. Indiana University Press,
    Bloomington.
7. Banner, A. 1968. Measurements of the particle velo-
    city and pressure of the ambient noise in a shallow
    bay. J. Acoust. Soc. Amer. 44(6):1741-1742.
8. Banner, A. 1972. Use of sound in predation by young
    lemon sharks, Negaprion brevirostris (Poey). Bull.
    Mar. Sci. 22(2):251-283.
9. Banner, A. and M. Hyatt. 1973. Effects of noise on
    eggs and larvae of two estuarine fishes. Trans.
    Amer. Fish. Soc. 102(1):134-136.
10. Brawn, V.M. 1961. Sound production by the cod
    (Gadus callarias L.). Behaviour 18:239-255.
11. Buerkle, U. 1968. Relation of pure tone thresholds to
    background noise level in the Atlantic cod (Gadus
    morhua). J. Fish. Res. Bd. Canada 25:1155-1160.
12. Buerkle, U. 1969. Auditory masking and the critical
    band in the Atlantic cod, Gadus morhua. J. Fish.
    Res. Bd. Canada 26(5):1113-1119.
13. Cahn, P.H.,W. Siler and J. Wodinsky. 1969. Acoustico-
    lateralis system of fishes: tests of pressure and
    particle-velocity sensitivity in grunts, Haemulon
    sciurus and Haemulon parrai. J. Acoust. Soc. Amer.
    46(6):1572-1578.
14. Chapman, C.J. 1973. Field studies of hearing in tele-
    ost fish. Helgolander wiss. Meeresunters 24:371-390.

15. Chapman, C.J. and A.D. Hawkins. 1973. A field study of hearing in the cod, Gadus morhua L. J. Comp. Physiol. 85:147-167.
16. Cummings, W.C. and P.O. Thompson. 1971. Underwater sounds from the blue whale, Balaenoptera musculus. J. Acoust. Soc. Amer. 50(4):1193-1198.
17. Cummings, W.C., B.D. Brahy and W.J. Herrnkind. 1964. The occurrence of underwater sounds of biological origin off the west coast of Bimini, Bahamas. pp. 27-43. In: W. Tavolga (ed.), Marine Bio-Acoustics. Pergamon Press, New York.
18. Diercks, K.J., T. Trochta and W.E. Evans. 1973. Delphinid sonar: measurement and analysis. J. Acoust. Soc. Amer. 54:200-204.
19. Dietz, F.T., J.S. Kahn and W.B. Birch. 1960. Non-random associations between shallow water ambient noise and tidal phase. J. Acoust. Soc. Amer. 32:915.
20. Enger, P.S., A.D. Hawkins, O. Sand and C.J. Chapman. 1973. Directional sensitivity of saccular microphonic potentials in the haddock. J. Exp. Biol. 59:524-533.
21. Evans, W.E. 1967. Vocalization among marine mammals. pp. 159-186. In: W.N. Tavolga (ed.), Marine Bio-Acoustics, Vol. 2, Pergamon Press, New York.
22. Evans, W.E. 1973. Echolocation by marine delphinids and one species of fresh-water dolphin. J. Acoust. Soc. Amer. 54:191-199.
23. Fay, R.R. 1974. Masking of tones by noise for the goldfish (Carassius auratus). J. Comp. Physiol. Psychol. 87(4):708-716.
24. Fish, J.F. and W.C. Cummings. 1972. A 50-dB increase in sustained ambient noise from fish (Cynoscion xanthulus). J. Acoust. Soc. Amer. 52(4):1266-1270.
25. Fish. J.F. and G.C. Offutt. 1972. Hearing thesholds from toadfish, Opsanus tau, measured in the laboratory and field. J. Acoust. Soc. Amer. 51:1318-1321.
26. Fish, M.P. 1964. Biological sources of sustained ambient sea noise. pp.175-194. In: W.N. Tavolga (ed.), Marine Bio-Acoustics. Pergamon Press, New York.
27. Fletcher, H. 1940. Auditory patterns. Rev. Mod. Phys. 12:47-65.
28. Fry, F.J., G. Kossoff, R.C. Eggleton and F. Dunn. 1970 Threshold ultrasonic dosages for structural changes in the mammalian brain. J. Acoust. Soc. Amer. 48(6):1413-1417.
29. Gray, G.A. and H.E. Winn. 1961. Reproductive ecology and sound production of the toadfish, Opsanus tau. Ecology 42(2):274-282.

30. Ha, S.J. 1968. Masking Effects On the Hearing of the Lane Snapper, Lutjanus synagris (Linnaeus). Thesis, University of Miami. 51 pp.

31. Ha, S.J. 1973. Aspects of Sound Communication in the Damselfish, Eupomacentrus partitus. Ph.D. Dissertation University of Miami. 78 pp.

32. Hall, J.D. and C.S. Johnson. 1971. Auditory thresholds of a killer whale Orcinus orca Linnaeus. J. Acoust. Soc. Amer. 4(1):515-517.

33. Hawkins, A.D. 1974. The sensitivity of fish to sounds. Oceanogr. Mar. Biol., Ann. Rev. 11:291-340.

34. Hawkins, A.D. and C.J. Chapman. 1975. Masked auditory thresholds in the cod, Gadus morhua L. J. Comp. Physiol. 103:209-226.

35. Hawkins, J.E. and S.S. Stevens. 1950. The masking of pure tones and of speech by white noise. J. Acoust. Soc. Amer. 22(1):6-13.

36. Heindsmann, T.E., R.H. Smith and A.D. Arneson. 1955. Effect of rain upon underwater noise levels. J. Acoust. Soc. Amer. 27:378-379.

37. Jacobs, D.W. and J.D. Hall. 1972. Thresholds of a freshwater dolphin, Inia geoffrensis Blainville. J. Acoust. Soc. Amer. 51(1):530-533.

38. Johnson, C.S. 1966. Auditory thresholds of the bottlenose porpoise (Tursiops truncatus, Montagu). U.S.Naval Ordinance Test Station Rep. (NOTS TP 4178).

39. Johnson, C.S. 1967. Sound detection thresholds in marine mammals. pp. 247-255. In: W.N. Tavolga (ed.), Marine Bio-acoustics, Vol. 2, Pergamon Press, New York.

40. Johnson, C.S. 1968. Masked tonal thresholds in the bottle-nosed porpoise. J. Acoust. Soc. Amer. 44: 956-967.

41. Knudsen, V.O., R.S. Alford and W. Emling. 1948. Underwater ambient noise. J. Mar. Res. 7:410-429.

42. Licklider, J. 1959. Three auditory theories. In: S. Koch (ed.), Psychology: a Study of a Science, Vol. 1, McGraw-Hill, New York.

43. Maniwa, Y. 1971, Effect of vessel noise in purse seining. pp. 294-296. In: H. Kristjonsson (ed.), Modern Fishing Gear of the World. Fishing News (Books) Ltd., London.

44. Miller, J.D., S.J. Rothenberg and D.H. Eldredge. 1971. Preliminary observations on the effects of exposure to noise for seven days on the hearing and inner ear of the chinchilla. J. Acoust. Soc. Amer. 50(4):1199-1203.

45. Møhl, B. 1968. Auditory sensitivity of the common seal in air and water. J. Auditory Res. 8:27-38.

46. Møhl, B. and S. Andersen. 1973. Echolocation: high frequency components in the click of the harbour porpoise (Phocoena ph. L (sic)). J. Acoust. Soc. Amer. 54(5):1368-1372.

47. Myrberg, A.A. Jr. 1972. Ethology of the Bicolor Damselfish, Eupomacentrus partitus (Pisces: Pomacentridae: A Comparative Analysis of Laboratory and Field Behaviour. Animal Behav. Monogr. 5(3): 197-283.

48. Myrberg, A.A. Jr. In press. Underwater sound: its effect on the behavior of sharks. In: R. Mathewson and T. Hodgson (eds.), Sensory Biology of Elasmobranch Fishes. U.S. Government Printing Office.

49. Myrberg, A.A. Jr. and J.Y. Spires. 1972. Sound discrimination by the bicolor damselfish, Eupomacentrus partitus. J. Exp. Biol. 57:727-735.

50. Myrberg, A.A. Jr. and J.Y. Spires. 1977. Comparative analysis of hearing among damselfishes of the genus, Eupomacentrus. Abstract. 57th Ann. Meeting, Amer. Soc. Ichthyologists and Herpetologists, Gainesville.

51. Myrberg, A.A. Jr., A. Banner and J.D. Richard. 1969. Shark attraction using a video-acoustic system. Mar. Biol. 2:264-276.

52. Myrberg, A.A. Jr., E.Spanier and S.J. Ha. In press. Temporal patterning in acoustical communication. In: E.S. Reese (ed.), Contrasts in Behavior. J. Wiley & Sons, New York.

53. Myrberg, A.A. Jr., S.J. Ha, S. Walewski and J.C. Banbury. 1972. Effectiveness of acoustic signals in attracting epipelagic sharks to an underwater sound source. Bull. Mar. Sci. 22(4):926-949.

54. Naval Ordinance Laboratory 1942.Measurement of background noise in the water at Cape Henry, Va., due to surf and marine life. N.O.L. Rept. No.594.

55. Nelson, D.R. 1967. Hearing thresholds, frequency discrimination and acoustic orientation in the lemon shark, Negaprion brevirostris (Poey). Bull. Mar. Sci. 17(3):741-768.

56. Nelson, D.R. and S.H. Gruber. 1963. Sharks: attraction by low-frequency sounds. Science 142:975-977.

57. Nelson, D.R. and R.H. Johnson. 1972. Acoustic attraction of Pacific reef sharks: effect of pulse intermittency and variability. J. Comp. Biochem. Physiol. 42A:85-95.

58. Nelson, D.R. and R.H. Johnson. 1970. Acoustic studies on sharks. Rangiroa Atoll, July, 1969. Tech. Rept. 2, Office of Naval Research Contr. No N00014-68-C-0318. 15pp.

59. Nelson, D.R. and R.H. Johnson. 1976. Some recent observations on acoustic attraction of Pacific reef sharks. pp. 229-237. In: A. Schuijf and A.D. Hawkins (eds.), Sound Reception in Fish. Elsevier, Amsterdam.

60. Nelson, D.R., R.H. Johnson and L.G. Waldrop. 1969. Responses in Bahamian sharks and groupers to low frequency, pulsed sounds. Bull. So. Calif. Acad. Sci. 68:131-137.

61. Norris, K.S. 1969. The echolocation of marine mammals. pp.391-423. In: S. Andersen (ed.), The Biology of Marine Mammals. Academic Press, New York.

62. Olsen, K. 1969. Directional hearing in cod (Gadus morhua). 8th I.F. Meeting, Lowestoft.

63. Olsen, K. 1971. Influence of vessel noise on behaviour of herring. pp. 291-293. In: H. Kristjonsson (ed.), Modern Fishing Gear of the World. Fishing News (Books) Ltd., London.

64. Piggott, C.L. 1964. Ambient sea noise at low frequencies in shallow water of the Scotian shelf. J. Acoust. Soc. Amer. 36(11):2152-2163.

65. Popper, A.N. and N.L. Clarke. 1976. The auditory system of the goldfish (Carassius auratus): effects of intense acoustic stimulation. Comp. Biochem. Physiol. 53A:11-18.

66. Popper, A.N. and R.R. Fay. 1973. Sound detection and processing by teleost fishes: a critical review. J. Acoust. Soc. Amer. 53:1515-1529.

67. Poulter, T.C. 1968. Marine mammals. pp.405-456. In: T. Sebeok (ed.), Animal Communication. Indiana University Press, Bloomington.

68. Ridgeway, S.H. and P.L. Joyce. 1975. Studies on seal brain by radiotelemetry. pp.81-91. In: K. Ronald and A.W. Mansfield (eds.), Biology of the Sea. Int. Counc. Explor. Sea, (I.C.E.S.) Rapp. and P.-V. Reun.

69. Sand, O. 1974. Directional sensitivity of microphonic potentials from the perch ear. J. Exp. Biol. 60:881-889.

70. Sand, O. and P.S. Enger. 1974. Possible mechanism for directional hearing and pitch discrimination in fish. pp. 223-242. In: J. Schwartzkopff (ed.), Mechanoreception. Rheinisch-Westfalische Akad. L. Wiss. 53.

71. Schevill , W.E., W.F. Watkins and C. Ray. 1963. Underwater sounds of pinnipeds. Science 141(3575):50-53.

72. Schuijf, A. 1974. Field Studies of Directional Hearing in Marine Teleosts. Ph.D. Dissertation, University of Utrecht. 119 pp.

73. Schuijf, A. and M.E. Siemelink. 1974. The ability of cod (Gadus morhua) to orient towards a sound source. Experientia 30:773-774.

74. Schuijf, A., J.W. Baretta and J.T. Wildschut. 1972. A field investigation on the description of sound direction in Labrus bergylta (Pisces: Perciformes). Neth. J. Zool. 22:81-104.

75. Schusterman, R.J. 1968. Experimental laboratory studies of pinniped behavior. pp. 87-171. In: R.J. Harrison, R.C. Hubbard, R.S. Peterson, C.E. Rice, and R.J. Schusterman (eds.), The Behavior and Physiology of Pinnipeds. Appleton-Century-Crofts, New York.

76. Schusterman, R.J. 1974. Auditory sensitivity of a Calfornia sea lion to airborne sounds. J. Acoust. Soc. Amer. 56(4):1248-1251.

77. Schusterman, R.J. and R.F. Balliet. 1969. Underwater barking by male sea lions (Zalophus californianus). Nature 222:1179-1181.

78. Schusterman, R.J., R.F. Balliet and James Nixon. 1972. Underwater audiogram of the California sea lion by the conditioned vocalization technique. J. Exp. Anal. Behav. 17:339-350.

79. Schusterman, R.J., R.F. Balliet and S. St. John. 1970. Vocal displays underwater by the gray seal, the harbor seal, and the stellar sea lion. Psychon. Sci. 18(5):303-305.

80. Sivian, L.J. and S.D. White. 1933. On minimum audibility fields. J. Acoust. Soc. Amer. 4:288-321.

81. Steinberg, J.C., W.C. Cummings, B.D. Brahy and J.Y. MacBain (Spires). 1965. Further bio-acoustic studies off the west coast of North Bimini, Bahamas. Bull. Mar. Sci. 15(4):942-963.

82. Tavolga, W.N. 1967. Masked auditory thresholds in teleost fishes. pp. 233-245. In: W.N. Tavolga (ed.), Marine Bio-Acoustics, Vol. 2, Pergamon Press, New York.

83. Tavolga, W.N. 1971. Sound production and detection. pp. 135-205. In: W.S. Hoar and D.J. Randall (eds.), Fish Physiology, Vol. 5, Academic Press, New York.

84. Tavolga, W.N. 1974a. Sensory parameters in communication among coral reef fishes. Mt. Sinai J. Med. 41(2):324-340.

85. Tavolga, W.N. 1974b. Signal/noise ratio and the critical band in fishes. J. Acoust. Soc. Amer. 55(6): 1323-1333.

86. Tavolga, W.N. and J. Wodinsky. 1963. Auditory
     capabilities in fishes. Bull. Amer. Mus. Nat. Hist.
     126(2):177-240.
87. Tavolga, W.N. and J. Wodinsky. 1965. Auditory
     capacities in fishes: threshold variability in the
     blue-striped grunt, Haemulon sciurus. Anim. Behav.
     13(2-3):301-311.
88. Terhune, J.M. and K. Ronald. 1971. The harp seal,
     Pagophilus groenlandicus(Erxleben, 1777). 10. The
     air audiogram. Can. J. Zool. 49:385-390.
89. Terhune, J.M. and K. Ronald. 1972. The harp seal,
     Pagophilus groenlandicus (Erxleben, 1777). 3. The
     underwater audiogram. Can. J. Zool. 50:565-569.
90. Terhune, J.M. and K. Ronald. 1975a. Underwater
     hearing sensitivity of two ringed seals (Pusa hispida)
     Can. J. Zool. 53:227-231.
91. Terhune, J.M. and K. Ronald. 1975b. Masked hearing
     thresholds of ringed seals. J. Acoust. Soc. Amer.
     58(2):515-516.
92. Urick, R.J. 1975. Principals of Underwater Sound.
     McGraw Hill, New York. 384 pp.
93. Watkins, W.A. 1974. Bandwidth limitations and ana-
     lysis of cetacean sounds, with comments on "Delphinid
     sonar: measurements and analysis" [K.J. Diercks,
     R.T. Trochta, and W.E. Evans, J. Acoust. Soc. Amer.
     54:200-204 (1973)]. J. Acoust. Soc. Amer. 55(4):849-
     853.
94. Wenz, G.M. 1962. Acoustic ambient noise in the ocean:
     spectra and sources. J. Acoust. Soc. Amer. 34(12):
     1936-1956.
95. Wenz, G.M. 1964. Curious noises and the sonic en-
     vironment in the ocean. pp. 101-119. In: W.N.
     Tavolga (ed.), Marine Bio-Acoustics, Pergamon Press,
     New York.
96. Wenz, G.M. 1972. Review of underwater acoustic re-
     search: noise. J. Acoust. Soc. Amer. 51(2):1010-
     1024.
97. Widener, M.W. 1967. Ambient-noise levels in selected
     shallow water of Miami, Florida. J. Acoust. Soc.
     Amer. 42(4):904-905.
98. Winn, H.E. 1967. Vocal facilitation and the bio-
     logical significance of toadfish sounds. pp. 283-
     303. In: W. N. Tavolga (ed.), Marine Bio-Acoustics.
     Pergamon Press, New York.
99. Winn, H.E. and P.J.Perkins. 1970. Sounds of the
     humpback whale. pp. 39-52. In: Proceedings of the
     seventh annual conference on biological sonar and
     diving animals. Stanford Research Institute, Menlo
     Park.

PRELIMINARY RESULTS OF THE EFFECTS OF NOISE
ON GESTATING FEMALE MICE AND THEIR PUPS

M.C. Busnel ET D. Molin

Laboratoire de Physiologie Acoustique
I N R A
Jouy-en-Josas, France

INTRODUCTION

For many years our laboratory has been studying the effect
of noise stresses on the physiology and behavior of various
animals. The present study follows the same line but concerns
the possible effect of a noise stress at a period of life when
the structure of the animal is most susceptible to change,
i.e., during the embryogenesis.

The triggering incentive for this research came from three
directions:

1) We were intrigued by noise studies on rats by Sackler and
    Weltman which showed that both sexes presented signs of
    stress, i.e, evidence of gonadal inhibition as well as
    increased adrenal function (16, 17, 18). Furthermore,
    in their studies of environmental stimulated-subway
    stress (noise, vibration and crowding), they reported low
    birth rates (19) and a significantly lower percent of
    male pups (unpublished results). The latter two findings
    were not confirmed in later more extensive studies (20).
2) The results of Arvay (1), Zondek and Tamari (24), Ward
    and al. (22) and principally of Geber and al. (6,7,8)
    on this subject had for many years made us wonder about
    the importance of a synergy of stresses during pregnancy.
3) The study carried out concurrently (Busnel and Granier-
    Deferre, unpublished data) on prenatal "memory" of a
    sound stimulus made us wonder about the importance of a
    prenatal event.

Our purpose here is not to review the studies on this subject which will be done in a later paper, but to present the
preliminary results of this long term project.

209

Our study is original in that:

1) While the majority of previous studies had been
   carried out on small samples of rats, we chose to
   use mice, which allows a greater sample size and its
   therefore easier to handle statistically. We have
   used 470♀ with 6 litters each, amounting to approximately 17,000 pups (see Table II).
2) We added two stresses (like Sackler and Weltman on
   the rat): noise plus a horizontal vibration which we
   will call "shaking". The shaking was accompanied
   by crowding (see Methods).
3) We used a deaf subline as well as a non deaf subline
   of the same mouse strain. The advantages of this
   technique will be explained in the "Methods" section.
4) We used a chronic stimulus (over 6 months exposure).
5) Instead of using a pure frequency or white noise,
   we used an urban noise to which many people are subjected daily, subway noise, chosen for eventual comparison of our results with those of Sackler and
   Weltman.

Although it is obvious that no direct correlation should
be drawn between mice and human beings, as J.G. Forsberg
(1975) said for endocrine effects "the similarity of the process by which the lesions in the mouse are produced and their
histologic similarity to the human lesions is striking".....
"the similarities between the lesions seen in mice and man...
should be more stressed than the differences" (5).

In our opinion, mice are not to be rejected for this
type of experiment. The objection of Kryter (13) who denies
the utility of rodents for all acoustic research, seems to
have been based on the fact that some strains of mice have
audiogenic seizures. This is a well known phenomenon, but
has nothing to do with research performed on genetic strains
where this specific sensitivity is not present.

We want to stress here, that the study described below
is preliminary as the difference between the deaf and non
deaf strain of GFF has not yet been measured, nor has that
between deaf and non deaf pups. Where ever parts of the
experiment are lacking it will be noted.

MATERIAL

A. Biological Material

Three strains of mice were selected:

A. Swiss albino Rb_3 : (normal hearing)*
B. G.F.F. +/+: (normal hearing)**
C. G.F.F. dn/dn: A deaf mutant obtained from University College, London from the GFF +/+ strain. The gene (dn) responsible for deafness is recessive.***

Prior to the experimentation, crosses were made between GFF dn/dn and GFF +/+ animals to obtain F_1 hybrids (GFF dn/+ which show normal hearing but which are carriers of the gene for deafness. Three groups of animals were subsequently used for our experimentation:

1. Swiss albino Rb_3 males were mated to Swiss albino females (normal hearing).
2. Male GFF dn/+ hybrids were mated to GFF dn/dn females, producing litters in which approx. 50% of the pups were deaf and 50% non-deaf, the former acting as controls for the litter.
3. Male GFF dn/dn (deaf) mated to GFF dn/+ (hearing hybrid) : again approx. 50% of the young were deaf and 50% non-deaf in the litter.

* Mouse News Letter 1969 n 21 (Companion issue) P. 42.
** Originated from the Glaxo Laboratory and obtained from University College, London
*** Deol (M.S.) and Kocher (W.): A new gene for deafness in the mouse, J. Heredity 1958 12:463-466. Ref: Mouse News Letter 1958 (18) 49.

The difference between group 2 and 3 is that the hearing pups of group 2 have hearing mothers while the hearing pups of group 3 have deaf mothers (see Table I).

TABLE I

Experimental breeding design (groups 2 and 3)

Females	Males	
	Hybrid GFF dn/+	Deaf GFF dn/dn
Hybrid GFF dn/+ (normal hearing)		Pups obtained: Deaf and non deaf (group 3)
Deaf GFF dn/dn	Pups obtained: Deaf and non- deaf - pups of deaf mothers (group 2)	

This experimental design may help in determining whether the potential effect of a sound stress is due to stress of the mother, or to stress of the foetus directly, or both.

In each group two-thirds of the animals were exposed to acoustic stress and the other third, acting as controls, received no such acoustic stimulation. The young of the a-coustically stressed animals were then compared with those of the controls.

B. Breeding Cages and Maintenance Conditions

Acoustic sources were placed in a room acoustically designed for our purposes (Figure 1) measuring 3, 6 m x 2, 4 m x 4 m. A circadian light program 12 h/12 h was provided by fluorescent "daylight" lamps with a luminosity per m 2 of 34,800 lux. Temperature is 25° C+ 1 and humidity 75 + 5.

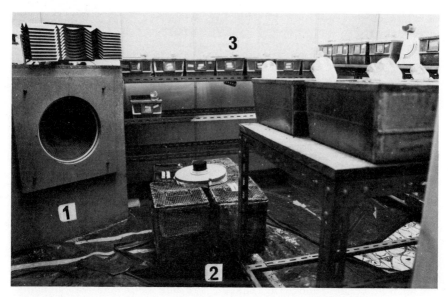

Figure 1
Experimental facilities in the sound proof room.
    (1) Loud speaker for medium and low frequencies.
    (2) Shaker supporting the cages for crowding and vibra-
        tions.
    (3) Some of the Breeding cages.

Figure 2
Details of shaker and cages.

Breeding cages measured 30 x 20 x 16 cm and contained wood
shavings. The animals received food and water "ad libitum"
except during periods of shaking stress. Their place in the
breeding room was changed every 2 weeks so as to insure
similar light and sound exposure.

Females were mated between 42 and 50 days of age with
60 day-old males and the couples formed were exposed to noise
from the day of mating.

According to the experimental conditions which will be
discussed later, the cages contained either 2♀ and 1♂ or 1♀
and 1♂ only.* The young were weaned at age 21 days.

C.   Noise Characteristics

The noise used was recorded on the inside of a Parisian
(R.A.T.P.) subway car with metal wheels. It represented a
typical journey (departures, stops, door openings and
closings) (Figure 3 - following page) at the actual acoustic
level to which human passengers are normally exposed : 105 ±
5 dB SPL. The recording lasted for one hour and was played
back at fixed intervals four times per day. The electroacoustic system used for the experimentation consisted of:

   1.   A Revox A 77 tape recorder (19 cm/s)
   2.   A Power amplifier (30 W)
   3.   A J.B. Lansing Type LE 15 loud speaker for higher
        frequencies (600 Hz - 14 kHz).

* During the summer of 1976, it became obvious that the
results obtained should be verified. There were too many
overlaps between the litters of the two♀. From September
1976 on the same experiment was repeated, using 15 cages
per mouse/strain, each containing one isolated couple.

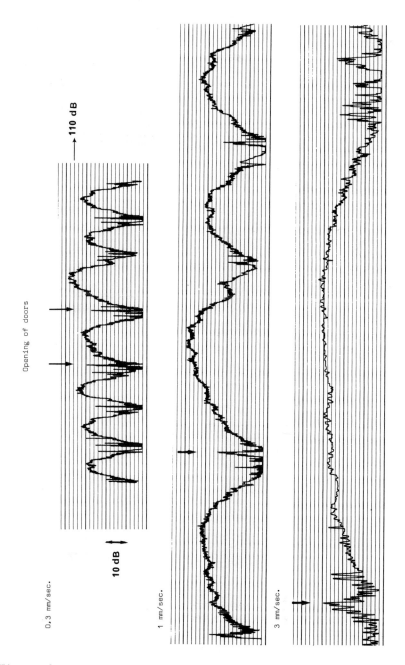

Figure 3
Sequential analysis at different speed of subway noise play-
back.

A "shaker" was used to simulate subway car vibrations. It was placed on a sheet of cork in which four special screened cages were embedded (30 cm x 22 cm x 18 cm) (Figure 2). 96 rhythmic movements per minute were thus produced whose horizontal amplitude was 2, 5 cm on either side of a central point (Figure 4).

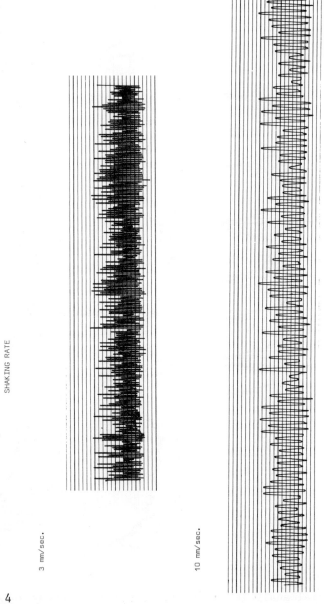

SHAKING RATE

3 mm/sec.

10 mm/sec.

Figure 4
Sequential analysis of the shaking rate, note irregularity of shaking amplitude.

A detailed analysis of the noise was carried out. A portion of the spectra of the recorded sound was taken from a magnetic tape recording and analyzed by a Bruël and Kjaer spectrograph (n°2305). The two loud-speakers described above worked together and were placed in an anechoic chamber of 5 x 4 m for a preliminary precise measurement; the same measurements were then repeated in the experimental breeding room which was specially designed to yield an acoustic insulation of the order of 40 dB SPL without reverberation. Care was taken to obtain measurements after the breeding cages were put into place, and observed with respect to the angle relative to the noise source in three different positions. The spectra differ with intensity variations on the order of $\pm$ 5 dB. The spectral density of the noise was calculated from a loop of the same recording by a I/3 octave Hewlett-Packard analyzer (Figure 5 and 6).

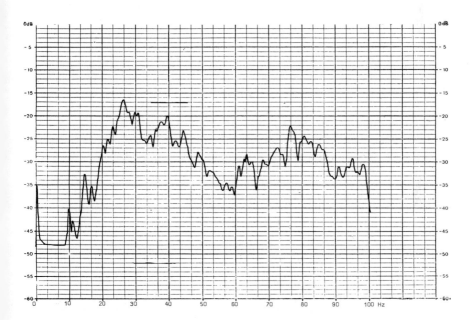

Figure 5
Average spectral analysis, of the Parisian subway detailed analysis up to 100 Hz.

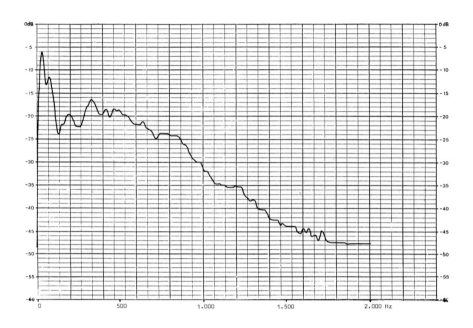

Figure 6

Average spectral analysis, of the Parisian subway analysed up to 2000 Hz.

According to the spectral analysis, the French subway emits a wide band of noise at the very low frequencies (between 15 and 40 Hz). These frequencies contained more energy (20 dB more) than the rest of the spectral bandwidth.

METHODS

1. Experimental procedure (see Table 2). Swiss Albino females (2♀ and 1♂ per cage) and GFF dn/dn females (2 GFF
   dn/dn/♀ and 1 GFF dn/+♂ per cage) were used as controls.
   At a later time, the same procedure was used with 1♀ and
   1♂/cage.
2. A similar group of mice was exposed in their breeding
   cages to four hours of noise, seven days a week. The
   recording lasted one hour and was played back from 7 to
   8 a.m., 11 a.m. to noon, 1 to 2 p.m.; during the week.
   The rhythm was purposely irregular on Saturday and Sunday
   in order to prevent too much habituation.

3. A third similar group of mice was exposed 5 days a week to the same noise but the females in addition were separated from males and litters (which remained in their breeding cages), crowded into wire cages (20 females/cage) and put on the shaking device during the mid-day noise exposure. The two hours of shaking were separated by one hour of silence without shaking during which time the animals remained grouped and received water ad libitum.

TABLE II

Different experimental treatment applied

Exp. N°	Animal Strain	Groups	Control Nb ♀	Noise Nb ♀	Noise + shaking + Crowding Nb ♀
I	Swiss Albino	2 ♀ + 1 ♂	60	40	40
II	Rb3	1 ♀ + 1 ♂	15	–	–
III		1 ♀ + 1 ♂	20	–	–
I	GFF	2 ♀ + 1 ♂	40	40	40
II	Deaf ♀ hearing ♂	1 ♀ + 1 ♂	15	–	15
III		1 ♀ + 1 ♂	25	20	15
I	GFF hearing ♀ deaf ♂	–	–	–	–
II		–	–	–	–
III		1 ♀ + 1 ♂	30	15	15
TOTAL			205	115	150

A. Quantified Observations of Animals

Breeding was begun in December 1975. All of the mothers were weighed three times a week, and daily during the 5 to 8 days preceding birth. The first five litters of each female were weighed daily* in some series, or on days 1, 7, 14 and 21 (day of weaning) in others.

* Because frequent weighing may in itself disturb the animals we have compared weighted and unweighted controls.

Before the birth of the 6th litter, each female was sacrificed by caesarian and the number, size and weight of each
embryo were noted, as well as any morphological macroabnor-
malities.

A comparison between the number of young after birth and
the number of embryo found by caesarian section allowed us
to obtain an indication of the rate of parental cannibalism.

DATA ANALYSIS*

In the six series receiving treatment, b and c and their
controls, the following characteristics were measured:

1. Fertility rate, inducing mortality of pups and parental
   cannibalism.
2. Regularity of irregularity of litter production.**
3. Sex ratio of the litters.
4. Growth curves: a) of pups b) of females.
5. Mortality of parents.
6. Malformation of pups or embryo.

THE STATISTICAL TESTS CONSISTED MAINLY OF:***

A. t Test (Student)
B. Comparison of 2 proportions by the corrected χ^2 (con-
   tingency table 2 x 2)
C. Comparison of two percentages.
   Normal:
   scores test

$$\mathcal{E} = \frac{(P_2 - P_1)}{\sqrt{Pq \left(\frac{1}{N_2} + \frac{I}{N_1} \right)}}$$

* Some of the results given must be considered incomplete
as all of the series are not yet terminated and it is therefore not possible to interpret them with all the reliability
one would wish. In complete results will be indicated.

** This is taken to mean all accidents which could be observed with respect to the normal 21 day cycle, i.e., uterine
resorption, precocious abortion, and irregular ovulation.
An example is given in Fig. 8.

*** We want to thank A. Bagady for performing the statistical analysis.

RESULTS

A.   Fertility Rate

The fertility rate was measured both by the member of pups at birth and by the difference between the number at birth and the number at weaning (mortality rate).

Of course, the number of litters in a given time will also intervene as a factor of fertility rate (see paragraph 2).

1.   *The Mean Number Of Pups*

The mean number of pups per litter for each is statistically similar at birth (Table 3 on the 1st series ($2♀ + 1♂$).

TABLE III

Mean number of young per litter at birth

	Litter	1	2	3	4	5	Average	Cesarian
Rb₃	Control	7,46	6,87	7,28	6,30	5,I0	6,60	7,42
	Noise	5,87	6,52	6,87	5,65	5,83	6,15	7,50
	Noise + shaking + crowding	5,83	6,01	6,00	6,5I	5,0I	5,87	7,75
	Undisturbed controls						8,I6	
GFF	Control	5,60	5,555	7,I5	5,5	5,82	5,92	7,40
	Noise	5,35	5,90	5,27	5,84	4,63	5 40	7,79
	Noise + Shaking + crowding	5,I6	5,6I	5,76	6,03	5,44	5,60	6,45
	Undisturbed controls	6,I4	6,36	6,59	6,38	5,94	6,38	

Table 3 shows the number of pups for all of the series, the mean per ♂ and the mean per ♀/litter - none of these results are significant. However when the % of pups disappearing between birth and weaning is considered the situation changes (see Table 4, last column). Frequent handling is probably responsible for the fact that our controls show 3 times more dead pups than undisturbed ones. Nevertheless the shaken and noise groups which show no differences between them, do show a statistically significant difference with the controls, 26, 5 and 28, 8% dead pups respectively, compared to 16, 3 for controls.

The GFF individuals submitted to noise + vibration + crowding (NVCr) show a similar % of neonatal death as hearing strain individuals receiving the same treatment (25, 4 for the GFF to 28, 8 for the Rb).

To explain that 15, 54 % of the "noise alone" deaf group are similar to the white control (16, 35%), it can be assumed that the subway noise frequency spectrum showing high intensitites in the low frequencies (15 to 40 Hz) and deaf mice having been proven to be sensitive to infrasounds* the results of the deaf group can be better understood. This needs further verification.

What can, however, be said is:

- that 25 to 29 % of lethality in the experimental pups is at any rate very high.
- that noise alone is as effective for the hearing strain as the triple stress (noise + shaking + crowding), and
- that, on the contrary, in the deaf strain the latter is more effective than noise alone.

* In another type of test (2,3) it was proved that the swimming time of deaf ♂ mice having been subjected to 2 h of infrasound is reduced by up to 50%, showing a definite effect of low frequencies on deaf animals.

2. *Rate Of Cannibalism*

The number of pups disappearing through cannibalism is however not truly shown by what can be inferred from table 4, since a number of females lost weight by approximatly the same amount as those giving birth to a live litter although no pups were seen. These pups must have been born but cannot be accounted for.

Fertility rate therefore must also include those ♀ which eat all pups of one litter but also ♀ which eat all pups of all their litters rather than one or two pups/litter.

Since there is no increase in the number of pups from the 2nd to the 5th litter (Table 4) it can be expected that the 6th would also have a similar number. Yet if one compares the number of foetuses found by caesarian section to the number of pups actually seen at birth, a small difference can be seen. It might therefore be inferred that there are more dead pups due to cannibalism than could be seen by direct observation, but this factor must be further investigated.

TABLE IV

% of pups disappearing before weaning

		% of pups disappearing before weaning			
	Treatment on ♀	Pups			% of disappearances of pups
		Total number	Mean pups/		
			♀	litter	
HEARING Rb	Undisturbed controls* 67 ♀	1894	28.27	5.65	4.65
	Controls 55 ♀	1425	25.9	5.18	16.35
	Noise 40 ♀	1176	26.5	6.6	26.5
	Noise + Shaking + Crowding 55 ♀	1610	29.3	5.85	28.8
DEAF GFF	Undisturbed controls* 42 ♀	1043	24.8	6.4	2.90
	Controls 35 ♀	1100	31.4	6.3	8.63
	Noise alone 40 ♀	1100	27.51	5.5	15.54
	Noise + Shaking + Crowding 35 ♀	928	26.9	5.8	25.43

* undisturbed control results were kindly furnished by Dr. Alice LEHMANN in our Laboratory

3.   Probable Uterine Resorption

When considering the weight curves of each   (see fig. 8) it can be seen a small increase in weight, which reaches a level of approximately one half of the normal level of weight at parturition, followed by a drop back to nonpregnant level. We have assumed that sudden drops corresponded to a miscarriage, and slow drops to uterine resorption.   The number of these "events" (measured in the Rb line as a first example) is highly significant both for the noise and noise + shaking groups (for 39   : controls = 5, 9%, noise alone II, 45% and noise + shaking + crowding 17, 6 %).

B.   Regularity of Litter Production

1.   Number Of Days Between Two Litters

Fig. 7 and 8 illustrate two types of birth rate, one very regular, the other very irregular, one keeping all her young, one eating most of her litters, it cannot be said that the regular type always occurs in the controls and the other always in the experimental animals, however, there were more irregular mothers in the latter than in the former (see below).

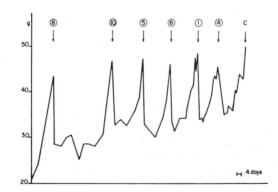

Figure 7
Example of female having rhythmical births.   Circle = Nb of pups.

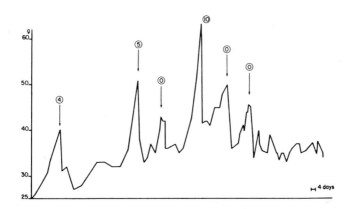

Figure 8
Example of female having unrhythmical births.   Circle =
Nb of pups.

There were more extremely long intervals between two
litters in the experimental groups (more than 41 days) and
more of what one might call "accidents" in the curve (the ♀
does not gain enough weight for a true litter and nothing is
found in the cage even though there is a sharp decrease of
body weight), as indicated in the preceeding section.

A ♀ who kills her litter or whose young are born dead,
and therefore does not lactate, will have a much greater
tendency to have the next litter exactly 21 days after the
preceeding one (or 20 days), while a lactating ♀ will be
more likely to have a slightly longer interval.   Therefore,
the mean time lapse between two successive litters is not in
itself a measure of a high fertility rate, but it is one of
the signs.

Table 5 (following page) gives the mean value for the
number of days between two gestations for each litter of each
♀ .   Considering the variability, the differences between
each litter are not significant.   Those exposed to noise a-
lone show no increase either, but in both strains (deaf and
non-deaf) the noise + shaken groups do show a general slow-
up.   This would be even more apparent if results from the
first litter (where there is very little possibility for
ovulation not to occur) were not taken into account.   When
comparing the 30 day mean which is found in the control with
the shaken Rb strain group (36, 1 days) and the deaf GFF
shaken group (44, 5 days) one sees highly significant in-

TABLE V

	Litter	1	2	3	4	5	$\bar{m}$	$\bar{m}$ without 1st litter
Rb	controls	31,92 ± 1.12	30,09 ± 0,98	30,94 ± 1,21	31,04 ± 1,07	27,8 ± 0,83	30,36 ± 0,47	30 ± 0,52
	noise alone	33,5 ± 2,06	36,82 **** ± 2,02	32,33 ± 1,53	29,22 ± 1,28	27,08 ± 0,98	31,93 * ± 0,78	31,51 * ± 0,83
	noise + shaking + crowding	29,025 ± 2,03	36,13 * ± 3,03	38,52 ** ± 3,31	42,86 ***** ± 2,08	33,925 *** ± 2,3	34,4 *** ± 1,62	36,23 ***** ± 1,78
GFF	controls	28,5 ± 1,52	32,91 ± 2,18	31,70 ± 2,02	30,525 ± 2,015	29,76 ± 1,78	30,7 ± 1,39	31,22 ± 1,5
	noise alone	33,5 * ± 2,06	36,82 ± 2,02	32,33 ± 1,5	29,22 ± 1,28	27,08 ± 0,98	31,79 ± 1,69	31,36 ± 2,11
	noise + shaking + crowding	35,76 *** ± 2,56	48,43 **** ± 4,6	45,18 ***** ± 2,61	45,08 **** ± 4,12	42,84 ***** ± 2,78	42,02 ***** ± 1,4	44,67 ***** ± 1,62

* p < 0,05   *** p < 0,01   ***** p < 0,0005
** p < 0,025   **** p < 0,0025

Mean length (in days) between two gestations.

crease. Therefore, the triple stress does reduce fertility through a lenghtening of the interval between two litters.

Another way of expressing the results would be counts of the % of gestations that last from 18 and 30 days, then the % that last from 31 and 42 days, and so on. One finds that the majority of control (68%) fall in the 18 to 30 days bracket, while all the experimental groups of non-deaf have a slightly higher percentage of longer gestations: 60 % between 18 and 30 days, 22% between 31 and 42 days.

2. *Regularity in Litter Production*

If instead of considering the mean value of the number of days between two litters, the number of actually irregular intervals are considered separately the following results was obtained.

TABLE VI

Lines	Treatment	Controls	Noise	Noise + Shaking
Rb_3	Number litters	80	73	71
	Number irregular	5	14	22
	%	0,6 25	19,18**	31***
GFF	Number litters	68	73	65
	Number irregular	9	20	27
	%	13	27, 4*	41,54***

*0,05 $<$ p $<$ 0,1   ** p $<$ 0,05   *** p $<$ 0,005

% Irregular intervals between litters $\geq$ 41 days.

The percentage is not significantly different between control and noise groups in the deaf strain, but very different from the shaken group, while it is significantly different between the experimental groups of the hearing mothers and the controls: the noise alone gives about 19% irregular and the shaking 31.

The same holds true if the percentage of having three or more regular intervals (for the five consecutive litters) is measured: 74% regular in the control, 55 in the noise group and 48 in the shaken. This is significantly different as from the control.

The irregularity of consecutive litters is therefore increased by the triple stress (NVCr) as well as is the interval between two consecutive litters. Noise alone has only an effect on the normal hearing group.

Irregular litter production is therefore greater in the experimental groups due both to longer intervals between litters, more irregular intervals and as shown in section 1c, a greater number of uterine resorption.

C. Sex Ratio

It had been thought that male embryo, being considered more fragile than female, might be less numerous and that therefore the number of males would be lower at birth than the number of females. This was not verified (see Table 7- following page).

TABLE VII

Sex Ratio (in %)

			1	2	3	4	5	Mean	Cesarian
Rb	Control	♂	60,14	53,49	56,51	60,08	53,69	56,95	56,29
		♀	40,91	46,84	43,49	39,92	46,31	43,36	43,71
	Noise alone	♂	53,37	52	55,71	58,08	50,87	53,99	47,04
		♀	46,63	48	43,38	41,92	49,71	45,90	32,22
	Noise + Shaking + Crowding	♂	52,51	50	51,91	50,88	64,14	53,31	41,94 *
		♀	47,49	50	48,09	49,12	35,86 *	46,69	37,10
GFF	Control	♂	47,06	45,63	55,04	49,50	40,48	48,17	43,24
		♀	52,94	54,37	44,96	50,50	59,52	51,83	49,32
	Noise alone	♂	53,19	56,83 *	47,45	52,20	48,84	51,77	50,94
		♀	47,87	43,17	52,55	47,80	51,16	48,34	47,55
	Noise + Shaking + Crowding	♂	48,23	45,65	49,40	42,14	52,78 *	47,47	
		♀	51,77	53,62	50,60	57,86	47,22	52,38	

*Statistically significant

WEIGHT OF THE PUPS

A preliminary analysis of results of the 1st series (in which 2♀ were kept together in the same cage) was done to determine if one litter's weight was influenced by the number of previous, simultaneous or subsequent litters.

The distribution for each litter's weight (Figure 9 - following page) is such that no correlation can be found for this factor and statistical analysis confirms this.

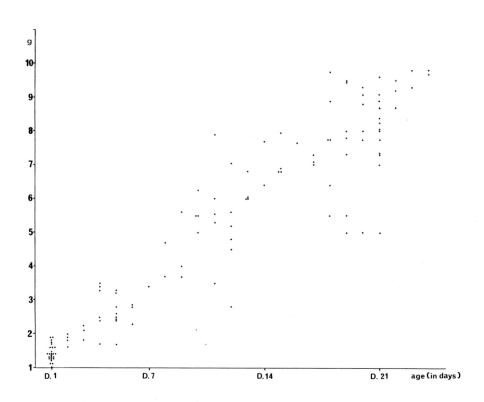

Figure 9
Mean weight for each litter of 30 females GFF dn/dn subjected
to the noise + shaking + crowding stress.

A second preliminary test on all litters of the 1st
series confirmed that pup weight was highly dependant on the
number of pups/litters (Figure 10 - following page). It thus
became necessary to study the mean number of pups/litter for
each series and their statistical distribution. Since it was
shown (see section 1a) that this mean number was not statisi-
cally different in the 3 groups, this too was disregarded in
the further analysis of litter weight.

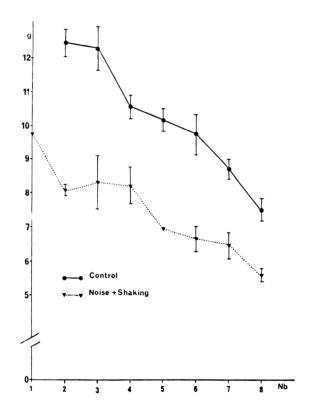

Figure 10
Weight at weaning as a function of number of pups per litter
(mean of 15 ♀)

    In the first series pups were weighed on specific days
of the week which meant at different ages (see Figure 9).
This technique was useful in following the general growth
curve. Once these results were obtained, the pups from later
series were systematically weighed at 1, 7, 14 and 21 days so
as to allow easier statistical analysis. However Figure 11
illustrates that the difference in weight at 21 days between
experimental and control groups is inversely proportional to
the number of pups in the litter (see also Table 8 last
column).

TABLE VIII

Nbre pups/litter Rb$_3$	Controls (Litter 1 to 5)		Noise + shaking + crowding (Litter 1 to 5)		P (t)	Diff. g
	mean weight (21 day) in g	St. deviation	mean weight (21 day) in g	St. deviation		
1						
2	12,47	± 0,44	8	± 0,17	< 0,0005	4.47
3	12,345	± 0,65	8,28	± 0,785	< 0,0025	4,06
4	10,61	+ 0,37	8,25	+ 0,54	< 0,005	2.36
5	10,21	+ 0,32	6,70	0	< 0,0005	3,51
6	9,8	+ 0,62	6,71	+ 0,37	< 0,0025	3.09
7	8,7	+ 0,29	6,49	+ 0,38	< 0,0005	2.21
8	7,48	+ 0,33	5,6	+ 0,22	< 0,0005	1.88
9	7,9					
10	5 9		4,775	+ 0,25		1,33
11	6	+ 0,51	5,45	+ 0,55		0,55

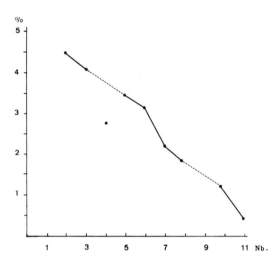

Figure 11
Mean difference between the weight/litter (on 15 ♀) as a
function of the number of pups/litter, notice the decrease
of stressfull effect on the pups weight with increase in
number/litter.

Further analysis therefore will have to take this into
account. As the weight of ♂ and ♀ pups are not significantly
different until 21 days, these results will not, in most cases,
be considered separately; measurements are in progress to
follow the weight of pups up to 42 days. At that age, of
course, the difference between   and   begins to be quite
marked.

Analyses have been carried out on each litter separately. An example of weight curves are given in Figure 12 for
the Rb line and 13 for the deaf GFF dn/dn. The results of
noise alone are not given for clarity of the figure (see
Figure 14 and 15 for analysis of that factor), it can be
seen from Figure 12 and 13:

1) That the general weight gain of GFF pups is smaller than
the general weight gain of Rb pups (the overall weight of the
females is also smaller).
2) That the difference between control and experimental pups
is quite similar from one litter to another.

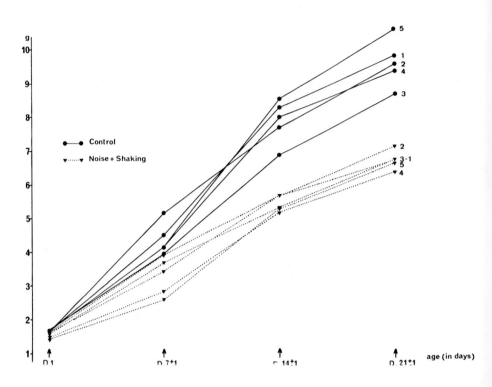

Figure 12
Effect of noise on growth of five consecutive litters of
non deaf mice (Rb$_3$).

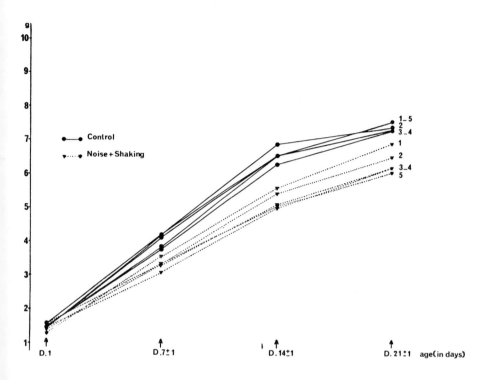

Figure 13
Effect of noise on growth of five consecutive litters of deaf
mice (GFF dn/dn).

3) That the difference between experimental and control
animals is smaller in the deaf GFF strain than in the hearing
groups.
4) That the slower growth intervenes more after 7 days than
before, and even more from 7 to 14 than from 14 to 21.   This
is the point at which controls show the sharpest increase in
weight (7 to 14 days) and the experimental animals seem to
lag in growth gain.

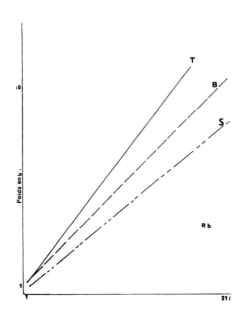

Figure 14
Mean difference between 1 and 21 days for the 3 treatments of hearing Rb$_3$ mice

T: Controls
B: Subjected to noise
S: Subjected to noise + shaking + crowding.

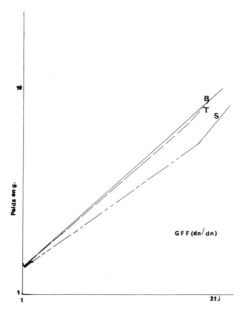

Figure 15
Mean difference between 1 and 21 days for the 3 treatments
of deaf GFF dn/dn mice.

T: Controls
B: Subjected to noise
S: Subjected to noise + shaking + crowding.

        Figure 14 and 15 show the mean of the 5 litters of all
females for each treatment in each series.   It can be seen
that:

1° - As in the two preceeding graphs, GFF pups as a group are
     smaller than Rb 3 pups.

2° - That the overall difference between control and experi-
     mental pup weight is smaller in GFF than in Rb.

3° - The main interest of these graphs is that:

     A - in the hearing Rb pups, noise alone has an effect in
         reducing pup weight, but less than the effect of the
         3 stresses (NVcr).

B - That in the deaf pups, as might be expected, the noise alone has no effect, and noise + shaking has an effect lower in real value but comparable in percentage to the results on pups of hearing mothers.

The regression coefficient of weight gain is highly significant for noise alone and NVCr in Rb line although it could not be measured for the GFF, the regression being irregular.

In conclusion, it can be said that the noise has a slight effect on hearing mother's pup weight, but no effect on the deaf ones, and that noise + vibrations + crowding (NVCr) has an effect on both strains of mothers. In the next section it will be seen that the same holds true for the weight of the mother herself.

WEIGHT OF FEMALES

It was thought that the mothers themselves could be disturbed by the noise and would therefore weight less at the time of parturition. In fact, this does not seem to apply. The mean weight attained just before birth by the pregnant ♀ was measured for each litter of 15 ♀ . The results are given Table 10 for the controls and noise + shaking + crowding for each group.

TABLE IX

Weight of ♀ at parturition for controls and ♀ submitted to noise + crowding + shaking.

		litter					$\overline{m}$ of the 5 litters
		I	2	3	4	5	
Rb$_B$	Control	46,4	49,9	5I,7	52,I	5I,I	50.225
	Experimental	44,I	**45,3	*47,4	49 3	**46,8	***46.6I
GFF (deaf)	Control	39,5	40,8	44	44	46,3	42.93
	Experimental	**36,9	39,8	42,4	43,23	***40,9	***40.65

* p < 0,I   ** p < 0,005   *** p < 0,0025 or more

It can be seen that the weight of the females is always lower in the experimental groups than in the control one but the differences are unevenly distributed among the different litters.

However, for the mean of all 5 litters, the difference although highly significant, represents only 7, 8 % of weight decrease in the Rb line and 5, 3 % in the GFF and we do not therefore feel that this factor can be considered very important for our study.

MORTALITY OF PARENTS

We have seen no sign of parental mortality induced by the applied stimulus. None of the ♂ died during the experiment. One or two of the ♀ in each series died, but these deaths were evenly scattered throughout the different series. No correlation could be found with any of the factors studied for such small number of details.

MALFORMATIONS OF THE EMBRYO

As indicated above, before delivery of the 6th litter, all were sacrificed and a caesarean section performed. All foetuses were examined for gross external and internal structural malformations, as well as any vacant uterine site indicative of foetal resorption.

As this part of the work is still in progress, we will give only qualitative results here. Even though we have up to now found many more malformations in the experimental groups we do not feel free to give numbers until all the series are terminated.

Only a few types of malformations have been found, of which some examples are given in Figures 16, 17, 18, and 19.

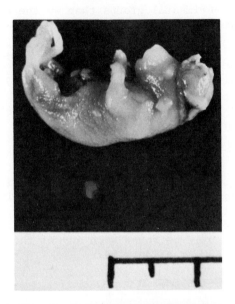

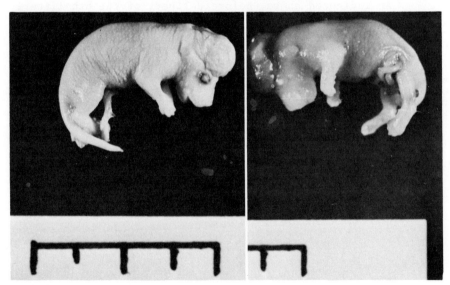

Figure 16
Examples of cranial malformations.

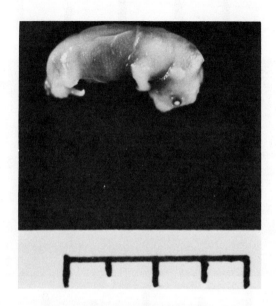

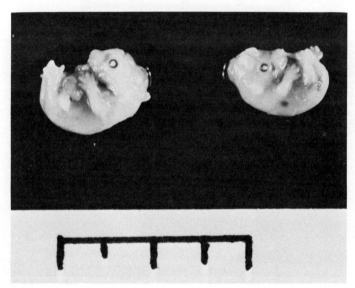

<u>Figure 17</u>

Examples of cranial malformations.

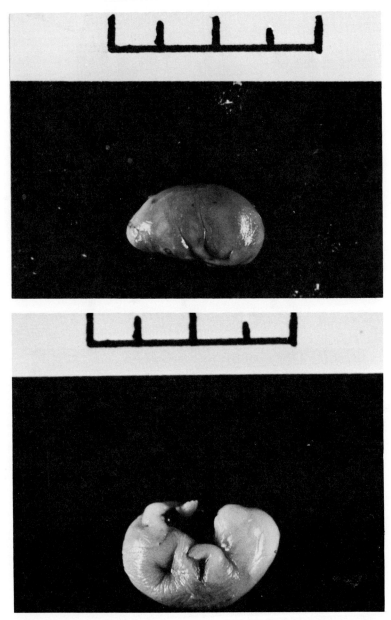

Figure 18
Examples of cranial malformations.

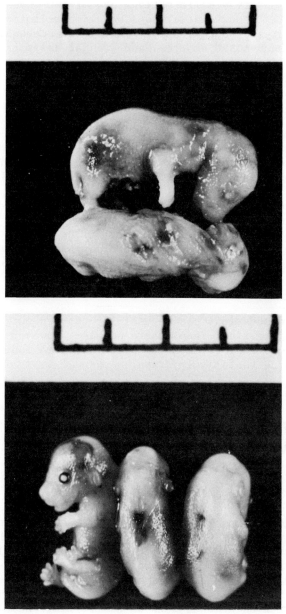

Figure 19
Examples of cranial and spinal malformations.

All observed malformations are of two types: cranial and spinal. We have never seen any limb malformation, no cleft palates. What we have seen is: exencephaly, anencephaly, snub-nosed embryos, and cranial assymetry. Also spinal malformations, some of which closely resemble spina bifida and uterine arrested development or growth retardation.

In the preliminary results, including the first series of treatments alone, we were able to count, for example, 1 % cranial malformation in the controls, 1 % in the noise alone, and 4 % in the noise / shaking groups. Spinal malformations were respectively 0, 1.85 and 6.85 % which gives for the cranial and spinal malformations, 1 % for control, 3 % for noise alone, and 6,85 % for the conjugated stress in the hearing strain.

The question of embryo malformation will result in a separate publication following project termination.

CONCLUSIONS

No discussions can be usefully carried out on general grounds since final results are not in. However a few results can be conisdered as complete and some conclusions can therefore be tentatively drawn.

1. There is no doubt that two stresses (or three if one considers crowding as a stress for these animals) are more effective than a single stress in affecting normal gestation.
Effect on gestation interval, gestational regularity, weight of pups and number of malformations is greater in the noise + shaking (or vibration) and crowding group than in the noise alone group.

2. The gestational disturbances attribuable to noise stress during gestation are real though discrete. They would probably pass unnoticed unless one specifically looks for them. It is therefore very important to study a large amount of data for proper statistical treatment and thus clarify all the uncertainties found in similar research on humans, such as controlling the belief that pilots and airport ground personnel had many more girls than boys. The negative results of Goerres' study (9) which relies on questionnaires of flying personnel are completely substantiated by our own.

3. Ishii in 1960 and Geber (1966) pointed out the importance
    of results on animals in this type of research. What we
    find very provocative is not that abnormalities, low
    weight, and uterine resorption can be produced by a
    sound stress; but that such a "normal" everyday event,
    as 4 hours of subway noise which is not noxious for the
    auditory organ, can, when given at a particularly sen-
    sitive period, produce such physiological effects in the
    mother as to be responsible for important changes in the
    embryo.

    Further study will have to elucidate the endocrino-
logical mechanism responsible for the observed perturbation
(probably those already described by Sackler and Weltman and
others, but whose origin must be pin-pointed). For this
purpose the use of genetically deaf animals should prove in-
valuable (study under way).

    Even though the mechanism must be further investigated,
it can already be inferred from the above results that the
mother is not such an effective barrier against noxious
stimuli as might have been thought, and that comparative
studies should be made of babies of women living in noisy
environments during pregnancy and others living in relative
silence.

    This is not because the noise in itself might be so
stressful, but because any added stress to women already
subjected to a high level of noise might produce physiological imbalances which the embryo is not able to neutra-
lize. Further research must also include stimulation at
specific time during pregnancy rather than during the whole
gestation period, so as to define the most sensitive period.

    Finally, the difference between deaf and hearing pups
of deaf or hearing mothers will undoubtly allow differentiation of effects due to stress of the mother or stress of
the foetuses themselves.

    Another point of possible discussion is to find out the
true relationship between the subjective effect of a certain
sound pressure on humans and on mice. It might turn out
that a given intensity is much more harmful to a mouse than
it can be to a human. Research on this point, although essential will be very difficult to carry out.

    We feel that the most important notion to be drawn
from these preliminary results is that cumulated non stressfull stimuli, can at certain sensitive periods become very
stressfull indeed.

ACKNOWLEDGEMENTS

We want to thank here Dr. Arthur M. Sackler (M.D.)
director of the Laboratories for Therapeutic Research, Research Institute of the Brooklyn College of Pharmacy, Long
Island University, for his support in carrying on the first
part of this research and the French Ministere de la Qualite
de la Vie et de l' Environment for his research grant N$^{\circ}$
75-94 which will allow us to finish the study.

REFERENCES

1. Arway, A.: Effects of exteroceptive stimuli on ferti-
    lity and their role in the genesis of malformations.
    Ciba. Found. Study Grp., 1967, 26:20-28.
2. Busnel, R.G., Busnel, M.C., and Lehmann, A.: Synergic
    effects of noise and stress on general behaviour.
    Life Sc., 1975, 16:131-137.
3. Busnel, R.C. and Lehmann, A.: Separation des effets
    psycho-physiologiques des infrasons et des sons par
    l'utilisation d'animaux genetiquement sourds. Revue
    d'Acoustique. (in press).
4. Chesser, R.K., Caldwell, R.S., and Harvey, M.J.: Ef-
    fects of noise on feral productions of mus musculus.
    Physiol. Zool., 1975, 48, n^{e}4, 323-325.
5. Forsberg, J.G.: Late effects in the vaginal and
    cervical epithelia after injections of diethylstil-
    bestrol into neonatal mice. Am. J. Obstet. Gynecol.
    1975, 121:101-104.
6. Geber, W.F.: Developmental effects of chronic
    maternal audiovisual stress on the rat fetus. J.
    Embryol. Exp. Morph., 1966, 16:1-16.
7. Geber, W.F.: Inhibition on fetal oestrogenesis by
    maternal noise stress. Fed. Proc., 1973, 32:2101-2104.
8. Geber, W.F., and Anderson, T.A.: Abnormal fetal
    growth in the albino rat and rabbit induced by
    maternal stress. Biol. Neonat., 1967, 209-217.
9. Goerres, H.P., and Gerbert, K.: Sex ratio in off-
    spring of pilots: a contribution to stress research.
    Aviat. Space and Environ. Med., August, 1976, 889-892.
10. Hollander, W.F.: Genetic Spina Bifida$_o$occulta in the
    mouse. Amer. J. of Anat., 1976, 146, n°2, 173-179.
11. Ishii, H., and Yokobori, K.: Experimental studies on
    teratogenic activity of noise stimulation. Gunma J.
    of Med. Sc., 1960, Vol. IX, n 1, 153.
12. Kimmel, C.A., Cook, R.O., and Staples, R.E.: Tera-
    togenic potential of noise in mice and rats. Toxicol.
    and Applied Pharmacol., 1976, 36:239-245.
13. Kryter, K.D.: Non-auditory effects of environment
    noise. Am. J. Public Health, 1972, 389-398.
14. Lieberman, L.S.: Prenatal auditory stimulation: ef-
    fects on developmental homeostasis, morphology and
    behavior in inbred mice. Diss. Abstr. Inter. Genetics.
    571B.
15. Sackler, A.M., and Weltman, A.S.: Effects of vibration
    on the endocrine system of male and female rats.,
    Aerospace Med., 1966, 37, 158-166.

16.   Sackler, A.M., and Weltman, A.S.: Endocrine and be-
       havioral aspects of intense auditory stress. Proceed.
       Psychophysiol.Biochem.Neuropharm. of Audit. Stress,
       1961, Nov. 6-9, C.N.R.S., I.N.R.A. ed. Paris.

17.   Sackler, A.M., Weltman,A.S., and Bradshaw, M., et.al.:
       Endocrine changes due to auditory stress. Acta Endocr.
       1959, 31:405-418.

18.   Sackler, A.M., Weltman, A.S., and Jurtshuk, P.: Endo-
       crine aspects of auditory stress. J. Aerospace Med.
       1960, 31:749.

19.   Sackler, A.M., Weltman, A.M., Pandhi, V., Johnson, L.:
       Effects of simulated-subway stress (SSS) on reproduc-
       tion and endocrine function of rats. Fed. Proc.,
       1975, 34:301.

20.   Sackler, A.M., Weltman, A.S., Johnson, L.: Effects
       of simulated-subway stress (SSS) on reproduction in
       rats. (Presented at the American Physiological Society
       28th Annual Fall Meeting, Oct. 9-14, 1977, Hollywood,
       Florida).

21.   Sinch, K.B. and Papineni, S.R.: Studies on the poly-
       cystic ovaries of rats under continuous auditory
       stress. Amer. J. of Obstet. and Gynec., 1970, 108, n°
       4, 557-564.

22.   Ward, C.O., Barletta, M.A., Kaye, T.: Teratogenic ef-
       fects of audiogenic stress in albino mice, J. Pharma-
       col. Sc., 1970, 59:1661-1662.

23.   Weltman, A.S., Sackler, A.M., Gennis, J.: Effects of
       handling on weight gains and endocrine organs in
       mature male rats, J. of Applied Physiol., 1961, 16,
       n°4, 587-588.

24.   Weltman, A.S., Sackler, A.M., Gennis, J., Steinglass,
       P.: Vibration effects on the endocrine function of
       rats, The Physiologist, 1966, 9:318.

25.   Zakem, H.B., Alliston, C.W.: The effects of noise
       level and elevated ambient temperatures upon selected
       reproductive traits in female swiss-webster mice,
       Lab.Anim.Sc., 1974, 24, n°3, 469-475.

26.   Zondek, B., Tamari, I.: Effects of auditory stimuli
       on reproduction, Ciba Found. Study Grp., 1967, 26:4-
       19.

DEINSECTIZATION OF STORED GRAIN BY HIGH
POWERED SOUND WAVES

A. J. ANDRIEU

Laboratoire de Physiologie Acoustique
I.N.R.A.-C.N.R.Z.
Jouy-en-Josas, France

F. FLEURAT-LESSARD

Station de Zoologie Agricole
I.N.R.A.
Pont-de-la-Maye, France

R. G. BUSNEL

Laboratoire de Physiologie Acoustique
I.N.R.A.-C.N.R.Z.
Jouy-en-Josas, France

INTRODUCTION

The pollution problems associated with chemical treatments of food products have always encouraged research into
alternative physical procedures. Among these, the use of
acoustic waves has long held the attention of investigators,
although attempts to apply it on an industrial scale have
continually met with failure.

The theory of acoustic wave utilization brings out the
fact that it is virtually impossible to obtain in-depth
energy diffusion in absorbing media. Whatever the wavelength used, the acoustic energy is absorbed by the surface
and is immediately degraded into heat. As a result, the
use of airborn ultrasonic frequencies for deinsectization is
infeasible. While surface temperature increase is considerable at high energies (iron shavings can even be ignited),
the insects lodged inside grain cannot be affected. It is
certain that at the focal point of an ultrasonic emitter

(a Hartmann whistle), any unprotected insect would quickly
find his body temperature reaching the coagulation point
for internal proteins. But the cost of the operation in
energy consumption is so high that it is unrealistic to consider its use on an industrial scale.

These negative conclusions, based on studies carried
out many years ago, could however perhaps be reconsidered
in the light of two new developments in high powered sound
equipment. One is a focalizer allowing power increases on
the order of 20 to 30 dB in a small space, and the other is
a generator developed in France capable of producing levels
of power never before reached. We have used a combination
of the two systems to reach even higher levels.

These new developments motivated us to accept this research problem in spite of our reservations on a theoretical
plane from the beginning and the slim odds for success.

The following experiments will be described succes-
sively:

1.   Laboratory trials using a focalizer, with the results
      of the physical measurements of diverse procedures at
      various frequencies and consecutive preliminary bio-
      logical observations.
2.   Trials with the high power generator.

STUDIES USING DISCRETE FREQUENCIES

In the first series of experiments, two bandwidths of
discrete sinusoidal frequencies were used: from 10 to 30
Hz, and a high-pitched 4 kHz frequency. To obtain high
sound intensity in these two bandwidths, different technological methods were used.

a)   Experiment 1 -- infrasonic frequencies (10 to 30 Hz).

Considering the wavelengths of such frequencies (from
11 to 33 m) and the noise intensity of conventional loud-
speakers (even those reputedly powerful), we designed the
following equipment for this bandwidth. A 600 m^3 locker
was divided into two parts by a central partition on which
the loudspeaker-acoustic wave generator was fixed. A cover
allowed hermetic sealing of the locker. The loudspeaker was
connected to a low frequency generator via a power amplifier.
Figure 1 represents the response curve of which such a
system and the maximum sound level attained with 15 V across
the terminals of the loudspeaker.

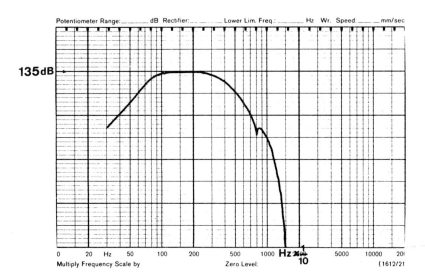

Figure 1 - Intensity and response curve obtained in a locker
used to generate very low and infrasonic frequencies.

A series of tests was conducted on Calandra granaria.
Test duration ranged from 1 to 10 minutes; frequencies
ranged from 10 to 30 Hz at an intensity of 135 dB.[1]   The
results were negative.

b)   Experiment 2 -- high frequency.

Several high efficiency high power compression chamber
loudspeakers now exist on the market. One of these is a
J.B. Lansing 375 model equipped with an acoustic lens allowing electroacoustic transmission of between 500 Hz and
10 kHz, with an amplitude on the order of 130 dB.

To be able to obtain a higher acoustic output, a paraboloid was used for amplification by focalization. The
paraboloid measured 80 cm in diameter and 40 cm in height,
with a focal-point-base distance of 10 cm. Figure 2 illustrates the response curve of the paraboloid, obtained
by placing a measurement microphone at its focal point.

[1]In this paper, all measurements are in S.P.L. dB.

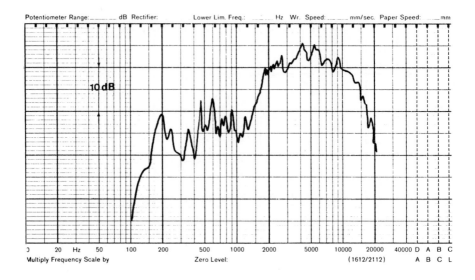

Potentiometer Range:_____ dB Rectifier:_____   Lower Lim. Freq.: ___ ___ Hz Wr. Speed: ___ ___ mm/sec. Paper Speed: ____mm

10 dB

Multiply Frequency Scale by   Zero Level: (1612/2112)   A   B   C   L

Figure 2 - Response curves of the paraboloid.

Maximum gain (24 dB) was highest at 4 kHz, and this frequency was consequently chosen for high intensity tests. We were thus able to obtain a sound intensity of 155 dB at the focal point of the paraboloid placed at a distance of one meter from the compression chamber. The chamber was associated with the acoustic lens which allowed collection of acoustic plane waves, thus producing a satisfactory focalization.

As in the previous experiment, tests were conducted on Calandra granaria and the results were again negative.

Considering the lack of success using discrete frequencies generated at the maximum possible intensity by conventional electroacoustic systems, we searched for alternative sources capable of producing much higher noise levels.

STUDY USING HIGH INTENSITY RANDOM NOISE

To obtain maximum sound pressure levels, pneumatic systems must be used. And to obtain random noise with a relatively wide bandwidth as well, use of a von Gierke type generator is advisable. Such facilities for high sound level trials are unique in France and recently imported, and were put at our disposal by the Societé Nationale Industrielle Aerospatiale.

Brief description[2]

The equipment includes:

-   a compressed air source obtained with a 500 CV compressor with a 2 m^3 air flow capacity.

-   a noise generator (random siren).

-   a plane pipe (wave-guide).

-   a reverberating chamber measuring 22 m^3.

-   air exhaust devices.

The noise generator essentially consists of a steel sheet metal box in which four pierced light alloy disks modulate the compressed air jet. The rotational speed of each disk is adjustable. A rotational speed modulation system allows discrete frequency generation to be avoided. In modifying the speed of rotation and modulation, it is possible within certain limits to obtain a given spectrum of generated random noise. Figures 3, 5, 7, and 9 illustrate diverse spectra obtained in the reverberating chamber.

To obtain a high sound level, we connected the paraboloid used in the preceding experiments. Adaptation of the random noise spectra made possible an intensity increase of 10 dB (raising the intensity up to 171 dB) at the focal point of the paraboloid placed in the reverberating chamber.

[2]
A complete description of the equipment was given by M.A. Rocache, engineer at S.N.I.A.S., in his dissertation defended at the University of Toulouse in December 1969. See also: M.A. Rocache, Revue d'Acoustique, Fr., 1971, 15, 148–152.

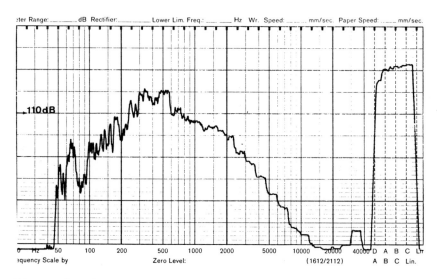

Figure 3 - Spectra obtained in the reverberating chamber from the S.N.I.A.S. siren at a global intensity of 120 dB.

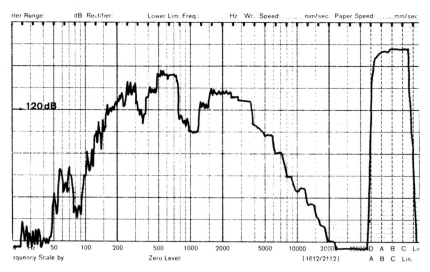

Figure 5 - Another type of spectra obtained in the reverberating chamber at a global intensity of 127 dB.

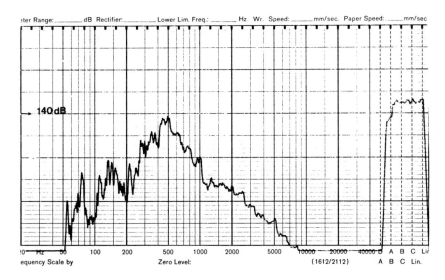

Figure 7 - Spectra obtained in the reverberating chamber at a global intensity of 143 dB.

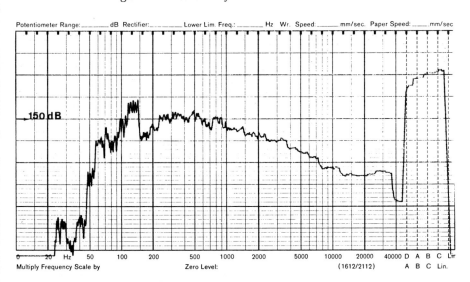

Figure 9 - Spectra obtained in the reverberating chamber at a global intensity of 161 dB, used for the trials.

Figures 4, 6, 8, and 10 illustrate spectra obtained using the paraboloid and correspond to Figures 3, 5, 7, and 9 respectively.

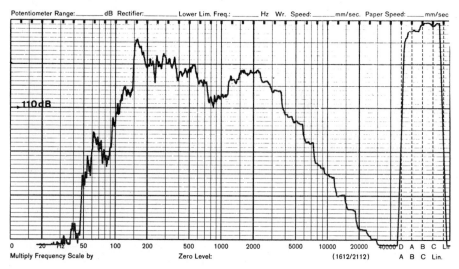

Figure 4 – Idem, at the focal point of the paraboloid.

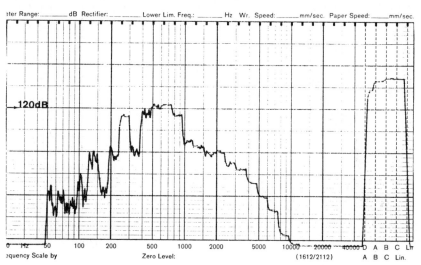

Figure 6 – Idem, at the focal point of the paraboloid.

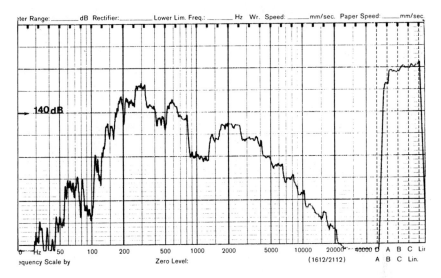

ter Range: _____ dB Rectifier: _____ Lower Lim. Freq.: _____ Hz Wr. Speed: _____ mm/sec. Paper Speed: _____ mm/sec.

140 dB

Hz 50   100   200   500   1000   2000   5000   10000   20000 40000 D A B C Lir

equency Scale by   Zero Level: (1612/2112)   A B C Lin.

Figure 8 - Idem, at the focal point of the paraboloid.

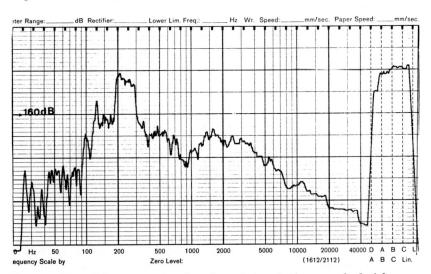

ter Range: _____ dB Rectifier: _____ Lower Lim. Freq.: _____ Hz Wr. Speed: _____ mm/sec. Paper Speed: _____ mm/sec.

160 dB

Hz 50   100   200   500   1000   2000   5000   10000   20000   40000 D A B C L

equency Scale by   Zero Level: (1612/2112)   A B C Lin.

Figure 10 - Idem, at the focal point of the paraboloid.

It was thus possible to carry out several tests on insect grain parasites.

Material Treated

a)   Insects

Treatments were carried out on batches of eggs, larvae, pupae, and adults of the Lepidoptera Ephestia kuhniella and larvae and adults of the Coleoptera Tribolium confusum, two species particularly noxious to hard wheat flour. Ephestia eggs were obtained from five couples of freshly emerged adults placed for four days in 70 g of SSSE flour. Adults were removed one day before treatment and replaced by young individuals, a process which allowed treatment of both eggs and adults in the same sample. Pupae and final instar larvae were sampled in a mass breeding culture on M.S. flour and placed in 70 g of fresh flour of the same quality to be treated together.

Tribolium adults ranged in age from 4 to 20 days. Larvae were in the final instar stage. Each instar stage was considered separately and each was composed of 50 individuals in 70 g of SSSE flour. Each insect group to be treated compreised 12 test trials of which four reference batches acted as controls.

b)   Conditioning

The groups destined for treatment were placed in cylindrical plastic boxes with large openings. Screens were put on the lids and on the two lateral windows. The lid was easily adjustable and held in place by adhesive tape.

Four tests were carried out with noise settings corresponding to Figures 9 and 10.

1.   On one batch of Ephestia and Tribolium adults in the reverberating chamber without the paraboloid at an intensity of 161 dB.
2.   On Ephestia (imago stage) placed in an acoustically transparent bottle at the focal point of the paraboloid at a global intensity of 171 dB.
3.   On Tribolium adults at the entrance of the generator plug at 180 dB, the highest intensity value that can be reached at the present time.
4.   On a set of groups of all types of insects placed in the reverberating chamber at an intensity of 159 dB.

The duration of all tests was limited to 15 seconds. Figure 11 summarizes the set of data relative to this last series of experiments on high intensity random noise.

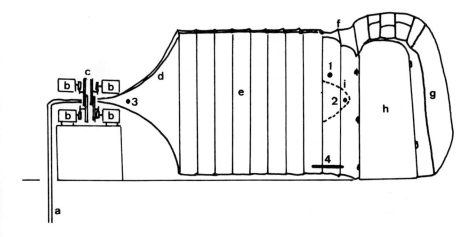

1, 2, 3, 4   Site of corresponding trials

a: Input of compressed air.   f:

b: Driving motor.   g: Reverberating chamber.

c: Pierced disk   h: Breeding culture.

d: i: Fine adjustment parabola

e: Plane pipe

Figure 11 - Schematic diagram of the high sound level equipment (S.N.I.A.S. siren used for experimental trials and localization of biological test sites).

RESULTS

1.   Rapid end of treatment observations.

Behavior of Coleoptera adults and Lepidoptera larvae was normal and their motor activity did not seem to be modified. During the trial at 180 dB, the breeding box was shattered, and yet Tribolium adults were found alive in the chamber. Two fragments of the box were subsequently soldered together (Figure 12). The distal wing parts of adult moths were broken and their scales were missing.

Figure 12 - Ephestia: adults before and after acoustic
          treatment.

2.   Subsequent observations

     Mortality observed at 48 h post-experimentation and
prolification observed four weeks after treatment are
grouped in Table 1 - 4.

CONCLUSIONS

     The negative results of the entomological observations
made following the diverse types of trials described in this
report discourage the adoption of sound wave exposure as a
deinsectization procedure. Furthermore, a study of operational costs in energy consumed alone, even at the trial
level, confirmed the inapplicability of such procedures
on an industrial scale, without even considering costs involved in large scale expansion of the facilities and
problems concerning subsequent safety measures for personnel
which must also be included.

RESULTS OF ENTOMOLOGICAL OBSERVATIONS[*]

1. Ephestia kuhniella: Adults and Eggs

Series Number	Sound Level (dB)	Exposure time	Number Insects Treated	Effects on adults 48 h after treatment			Effects on eggs Population size 4 weeks after treatment
				Dead	Alive	% mortality - adults	
1	-	0	10	10	0		+ +
2	-	-	10	10	0		+ + +
3	-	-	10	10	0		+ + +
4	-	-	10	3	7		+ + +
MEAN TOTAL			40	33	7	82.5%	
1	161	15 s	10	6	4		+ + +
2	"	"	10	3	7		+ +
3	"	"	8	5	3		+ +
4	"	"	10	3	7		+
5	"	"	10	6	4		+ + +
6	"	"	10	5	5		+
7	"	"	10	4	6		+ + +
MEAN TOTAL			68	32	36	47%	
1	170	15 s	10	8	2		+
MEAN						80%	

+ Low population   + + medium population   + + + high population

[*] Observations carried out by the Station de Zoologie of the I. N. R. A. at Pont de la Maye.

2. Ephestia kuhniella: Larvae and Pupae

Series Number	Sound Level (dB)	Exposure Time	Insects Treated	Emergence at 5 weeks after treatment	Population Size
1	-	-	40	39	+ +
2	-	-	40	38	+ + +
3	-	-	40	36	+ + +
4	-	-	40	36	+ +
MEAN TOTAL				149 93.1%	
1	161	15 s	40	34	+ + +
2	"	"	40	39	+ + +
3	"	"	40	38	+ + +
4	"	"	40	32	+
5	"	"	40	32	+
6	"	"	40	36	+ + +
7	"	"	40	38	+ +
8	"	"	40	39	+ + +
MEAN TOTAL				288 90.0%	

3. <u>Tribolium confusum</u>: Adults

Series Number	Sound Level (dB)	Number Insects Treated	Exposure Time	OBSERVATIONS 48 h after treatment			4 weeks after treatment		
				Dead	Alive	% Mort.	Dead	Alive	% Mort.
1	-	47	0	2	45	4.2	16	31	34
2	-	50		0	50	0	14	36	28
3	-	50		0	50	0	19	31	28
4	-	49		0	49	0	11	38	28.6
MEAN TOTAL		196		2	194	1%	60	136	30.6%
1	161	50	15 s	1	49	2	9	41	
2		50		-	50	-	8	42	
3		50		-	50	-	4	46	
4		50		-	50	-	14	36	
5		50		-	50	-	2	48	
6		50		-	50	-	15	35	
7		50		-	49	-	22	27	
MEAN TOTAL		349		1	348	0.3%	74	275	21.2%
1	180	50	15 s	Box broken, insects scattered (alive but lost).					

4. <u>Tribolium confusum</u>: Larvae (final instar)

Series Number	Sound Level (dB)	Exposure Time	Number Insects Treated	OBSERVATIONS					
				48 h after treatment			4 weeks after treatment		
				Dead	Alive	% Mort.	Dead	Alive	% Mort.
1	-	0	50	-	50	-	30	20	60
2	-	0	50	-	50	-	33	17	66
3	-	0	48	3	45	6.25	47		
4	-	0	50	2	48	4.2	40		100
MEAN TOTAL			198	5	193	2.5	160	38	80.1%
1	161	15 s	50		50		49	1	
2			52		52		39	13	
3			50	2	48	4	30	20	
4			50		50		11	39	
5			50	3	47	6	43	7	
6			50		50		30	20	
7			53	1	52	2	16	37	
8			50	2	48	4	32	18	
MEAN TOTAL			405	8	397	2%	250	155	61.7%

264

ACKNOWLEDGEMENTS

We would like to thank the director and collaborators of the Station de Zoologie Agricole of the Institut National de Recherche Agronomique for the kind welcome they gave us and for the entomological work that they undertook on the results of the trials reported here.

We also wish to thank the director of the Laboratoire Acoustique of the S.N.I.A.S. in Toulouse for the experimentation that he allowed us to carry out using his facilities and the help of his personnel.

QUANTIFYING THE ACOUSTIC DOSE WHEN DETERMINING THE
EFFECTS OF NOISE ON WILDLIFE

Robin T. Harrison

Forest Service, U.S. Department of Agriculture
Equipment Development Center
San Dimas, California

INTRODUCTION

A review of the literature on the effects of noise on
wildlife reveals that the acoustic levels and dose to which
test wildlife have been subjected   are not well quantified
(U.S. Environmental Protection Agency 1971). In some cases,
the noises of interest are described only by their source.
In others, simply an A- or C-weighted level is given. Very
few papers discuss the temporal patterns of noise and their
effects on wildlife, since sound level measurements at a
"wild" animal's ear can not be obtained in a conventional
manner. Nevertheless, a fairly precise picture of the spectral and temporal pattern that exists throughout a habitat
area can be predicted, using the method described in this
paper. This information is required to assess accurately
the environmental impact of noisy activities on wildlife.

The method, adopted from one developed to predict aural
detectability of military vehicles, utilizes a program that
can be run on a desk-top computer. The program uses source
spectral and temporal characteristics as input data, as
well as readily atmospheric parameters, and terrain and
vegetative parameters that can be fairly easily estimated.
Also needed are the distance and time distribution of noise
source from, and the hearing characteristics of, the subject
animal, and noise statistics of the background sound in the
habitat area. The program has been used with some success
under actual field conditions to predict the acoustic impact
of off-road vehicles on human forest users. This paper de-
scribes the computer program as it currently exists, and

267

presents a proposed pilot project to verify its applicability to large mammals.

The Forest Service, U.S. Department of Agriculture, is charged with the regulation of the environmental impact of each aspect of man's activities on lands that have been set aside for wildlife, wood, water, forage, and recreation resource management. For this reason, we believe the effects of sound must be studied as a separate phenomenon when considering how man's encroachment into the forest affects wildlife.

PREDICTIONS OF SPECTRAL AND TEMPORAL PATTERNS OF SOUND PRESSURE LEVEL

At the Forest Service's Equipment Development Center in San Dimas, California, an interactive computer program has been devised that can be run on a desk-top computer. (We use the Wang 2200). The program is an adaptation of one designed to predict the aural detectability of combat vehicles (Fidell and Bishop, 1974, Horonjeff and Fidell, 1977). Its output is a matrix of temporal and spectral distribution of sound pressure level at a hearer's location caused by various "encroaching" noise sources (Figure 1). A second part of the program predicts the distance from the encroaching source that a human hearer must be removed for the source to be just inaudible. We also propose to develop a program predicting the distance that an animal hearer would have to be removed.

The main input parameters needed to predict the temporal and spectral patterns of sound at the hearer's (animal or human) location are:

Time/frequency/level matrix of noise source at a known distance (assumed to be in the far field for this program)

Atmospheric conditions

Terrain

Ground impedance

Foliage and barriers (if any) between source and potential hearer

Actual distance between source and potential hearer

SITE 7
SAMPLE SIZE =   250
I=0·1

FREQ	MAX	MIN	SIG	LEQ	LEQ5	L1	L3	L5	L10	L20
50	49.4	33.4	2.2	42.3	42.1	47.0	46.0	45.4	44.8	43.6
63	47.4	36.4	2.0	41.4	41.2	45.9	44.8	44.2	43.4	42.6
80	46.4	36.4	1.9	41.9	41.7	45.6	45.1	44.8	44.1	43.1
100	44.4	34.4	1.7	40.1	40.0	43.9	43.2	42.7	42.0	41.2
125	39.4	30.4	1.5	35.3	35.2	38.6	38.1	37.8	37.1	36.3
160	36.4	26.4	1.3	31.6	31.5	35.2	34.0	33.5	33.1	32.5
200	35.4	26.4	1.3	30.9	30.8	34.2	33.2	32.9	32.3	31.8
250	36.4	28.4	1.1	32.2	32.1	34.9	34.2	33.9	33.4	33.0
315	36.4	29.4	1.1	33.0	32.9	35.6	35.1	34.8	34.3	33.8
400	38.4	31.4	0.9	34.6	34.5	36.6	36.2	36.1	35.6	35.2
500	37.4	32.4	0.8	35.6	35.6	37.3	37.2	37.1	36.7	36.3
630	38.4	33.4	0.8	35.6	35.6	37.4	37.2	37.0	36.6	36.2
800	38.4	34.4	0.7	36.2	36.2	37.9	37.4	37.3	37.2	36.9
1000	38.4	34.4	0.6	36.0	35.9	37.4	37.3	37.1	36.9	36.4
1250	38.4	34.4	0.6	36.5	36.5	38.1	37.5	37.4	37.3	37.1
1600	38.4	35.4	0.5	37.3	37.3	38.4	38.3	38.3	38.1	37.9
2000	38.4	35.4	0.4	37.0	37.0	38.3	38.1	38.0	37.5	37.3
2500	37.4	34.4	0.5	36.2	36.1	37.4	37.3	37.2	37.0	36.6
3150	36.4	33.4	0.5	34.5	34.5	35.4	35.4	35.3	35.2	35.1
4000	33.4	30.4	0.3	32.0	32.0	33.3	33.1	32.9	32.5	32.3
5000	30.4	26.4	0.3	28.0	28.0	29.3	29.1	28.9	28.4	28.3
6300	28.4	23.4	0.7	24.6	24.5	27.2	26.0	25.4	25.3	25.1
8000	25.4	17.4	0.8	19.0	19.0	23.9	20.5	20.0	19.4	19.2
10000	21.4	11.4	1.0	13.3	13.2	19.6	15.2	14.5	13.6	13.3
OA	54.9	46.5	0.0	50.4	50.3	53.6	52.9	52.6	52.0	51.3
DBA	48.4	44.6	0.0	46.5	46.5	47.9	47.7	47.6	47.3	47.0

FREQ	L30	L40	L50	L60	L70	L80	L90	L95	L97	L99
50	42.9	42.2	41.6	40.9	40.3	39.7	39.0	38.5	37.8	36.7
63	42.0	41.5	40.9	40.4	39.8	39.2	38.3	37.6	37.3	36.7
80	42.5	42.0	41.5	41.0	40.5	39.8	39.0	38.2	37.7	37.0
100	40.5	40.1	39.7	39.4	39.0	38.5	37.7	36.9	36.6	35.6
125	35.8	35.4	35.0	34.7	34.2	33.7	33.0	32.5	32.1	31.6
160	32.1	31.7	31.3	31.0	30.8	30.5	29.7	29.1	28.7	27.9
200	31.3	31.0	30.7	30.4	30.0	29.6	29.1	28.6	28.5	27.7
250	32.6	32.3	32.0	31.8	31.6	31.2	30.6	30.1	29.8	29.5
315	33.3	33.1	32.8	32.5	32.2	31.9	31.5	31.2	30.9	30.5
400	35.0	34.7	34.5	34.2	34.0	33.7	33.4	32.9	32.7	32.4
500	36.0	35.8	35.5	35.3	35.0	34.7	34.5	34.2	33.8	33.5
630	36.0	35.8	35.5	35.3	35.0	34.8	34.5	34.2	33.9	33.6
800	36.6	36.3	36.1	36.0	35.8	35.6	35.4	34.9	34.7	34.5
1000	36.2	36.1	35.9	35.8	35.6	35.5	35.0	34.7	34.6	34.5
1250	36.9	36.7	36.5	36.3	36.0	35.8	35.6	35.5	35.4	34.8
1600	37.6	37.3	37.2	37.0	36.9	36.7	36.5	36.4	36.4	35.8
2000	37.2	37.1	36.9	36.8	36.7	36.6	36.4	36.2	35.9	35.6
2500	36.3	36.2	36.1	35.9	35.8	35.6	35.5	35.4	35.4	34.9
3150	34.9	34.7	34.6	34.4	34.1	33.9	33.6	33.5	33.5	33.4
4000	32.2	32.1	32.0	31.8	31.7	31.6	31.5	31.5	31.4	31.4
5000	28.2	28.1	28.0	27.8	27.7	27.6	27.5	27.4	27.4	27.2
6300	24.9	24.7	24.5	24.3	24.1	23.8	23.6	23.5	23.5	23.4
8000	19.1	19.0	18.9	18.7	18.6	18.5	18.1	17.8	17.6	17.5
10000	13.2	13.1	13.0	12.8	12.7	12.6	12.5	12.4	12.4	12.2
OA	50.8	50.4	50.1	49.7	49.3	48.9	48.4	48.0	47.7	47.2
DBA	46.8	46.6	46.4	46.2	46.1	45.9	45.6	45.5	45.3	45.0

Figure 1 - Matrix of temporal and spectral distribution of
S.P.L. at hearer by "encroaching" noise sources.

Additionally, the spectrum of the background at the hearer's location must be known. While the background spectrum at any location is a function of time, our measurements have shown that for a time period of 1 or 2 hours (barring major ambience changes--such as thunderstorms) the background levels at the frequencies of most interest are fairly constant at any location. Then, a single background spectrum can be used and L_{50}, or the mean spectrum, appears to be the best choice (Harrison 1975). Backgrounds do vary greatly from place to place, even within a small geographical area.

While this computer program is discussed as if it deals with a single noise source as it sounds to a single hearer (receiver), the technique described has been expanded to multiple sources and multiple receivers by Horonjeff and Fidell (1977). The program is completely interactive-that is, it asks the user for each "next step" during the input process. It also, hopefully, only asks for information the user is likely to have; i.e., one-third octave band analyses, mean wind speeds, etc.

Air Absorption

The energy extracted from sound waves is estimated using the method of the Society of Automotive Engineers (SAE 1964). This shows good agreement between the theoretical calculations of Kneser (1933) and Piercy (1969). Both classical absorption and vibrational relaxation of nitrogen are accounted for. The model for air absorption per wavelength at 20°C and approximately 40 percent relative humidity is shown in Figure 2.

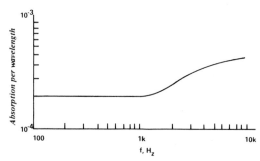

Figure 2 - Atmospheric absorption.

Effects of Atmospheric Turbulence

     Turbulence causes fluctuations in the intensity of
sound waves as they pass through the atmosphere. While
detailed knowledge of the scale and magnitude of turbulence
in the atmosphere will not be known to the program user,
these quantities are related to the mean wind velocity.
Fluctuations in attenuation are likely to be higher during
periods of high wind. Surface conditions may influence the
uniformity of wind and temperature gradients close to the
ground and will contribute to the fluctuations in atmospheric stability, and thus to variations in intensity.
These fluctuations are likely to be greater when both source
and hearer are on the ground than when propagation is from
an airborne source to the ground (Ingard and Maling 1963).
Fluctuations in broadband signals are significantly less than
those seen with pure tones. Corrections for fluctuations
in intensity due to turbulence are incorporated in the
program.

Scattering

     Non-uniformities in the atmosphere cause a negligible
scattering loss as a practical matter, particularly with
sources that are largely nondirectional (such as motor
vehicles, road construction, etc.). Atmospheric scattering
is ignored in the propagation model. Scattering effects
are important, however, in limiting the attenuation that can
be achieved by barriers or ground-plane reflections, and
are considered by the program.

Ground Plane Effects

     The effects of the ground plane on propagation over
level surfaces at short and intermediate distances is estimated following the methods of the SAE (1974). The basic
geometry of the situation is outlined in Figure 3. The
source is assumed to be a point source, which produces
stationary and random noise; the hearer is assumed to be in
the far field. Atmospheric absorption is ignored, since
the difference in direct-path length and total reflected-
path length is small compared to the length of each path.
The atmosphere is assumed to be still, isothermal, and homogeneous. Ray acoustics are assumed to be applicable, i.e.,
specular reflection is assumed and the concept of an image
source applied.

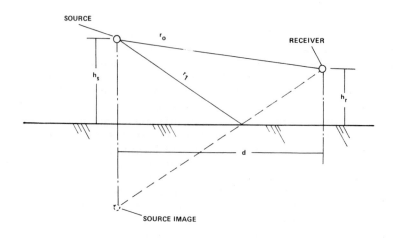

r_o = Length of Direct Sound Rays
r_1 = Length of Reflected Sound Rays
h_s = Height of Source Above the Ground
h_r = Height of Received Above the Ground
$r = r_1 - r_o$ = Difference Between Reflected and Direct Path Length

Figure 3 - Geometry for calculating ground plane effect.

In real life, of course, the ground is not a perfect
reflector and so this model is modified by incorporating
the impedance of the ground plane. Typical impedances for
three categories of surface--grassland, asphalt and con-
crete--have been placed in the computer's memory. The effect
of a finite impedance on excess attenuation in the 500-Hz,
one-third octave band is shown in Figure 4 (Sutherland 1973).

Wind and Temperature Gradients;
Upwind and Downwind Propagation

Refraction occurs whenever a sound wave is propagated
through an atmosphere which has a gradient in the velocity
of sound. In our case, the velocity gradient is brought
about by non-uniformity in the wind velocity or temperature,
or in both. The simplified model used in the program makes
gross assumptions about the shape and magnitude of wind and
temperature gradients because, in practice, detailed information necessary for more sophisticated predictions is
almost never available. The model is based, in the main,
upon data and analyses of Ingard (1969), Wiener and Keast
(1959), and Delany (1969). The wind gradient is assumed to

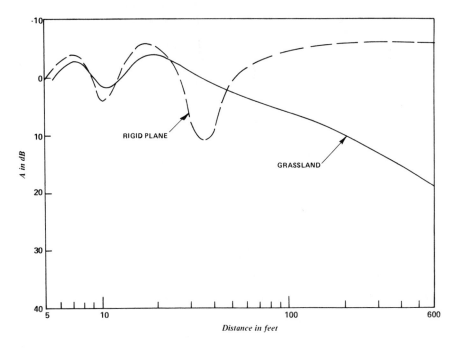

Figure 4 – Excess ground attenuation in 500 Hz one-third
octave band for grassland-type surfaces and for
rigid plane.

be of the form:

$$U = k \log \frac{h}{h_o} ,$$

where U is the wind velocity gradient, h is the height, and
h_o is a roughness parameter and k is a constant. The
temperature gradient is assumed to be of the form

$$\frac{dT}{dh} = -\frac{g}{R}\frac{\gamma-1}{\gamma} = -0.1°C/100 \text{ m,}$$

where g is the acceleration of gravity, R is the gas constant, and γ is the ratio of specific heats for air.

Figure 5a shows the effects of wind and temperature
gradients. Both wind and temperature gradients can result
in shadow zones; areas in which no sound waves may theore-
tically penetrate and are, therefore, areas of high attenua-

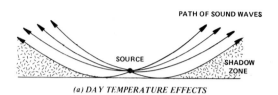

(a) DAY TEMPERATURE EFFECTS

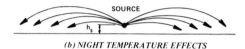

(b) NIGHT TEMPERATURE EFFECTS

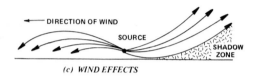

(c) WIND EFFECTS

Figure 5a - Effect of temperature and wind in forming sound shadow zones.

tion. Shadow zones are shown in Figure 5b for typical combinations of daytime and nighttime wind and temperature gradients (Rudnick 1957). The effectiveness of the shadow zone varies with both frequency and distance. The theoretical model for the formation of these shadow zones is very complex, but simplified equations have been developed by Fidell and Bishop (1974) based on studies by Ingard (1953), Wiener and Keast (1959), and Delany (1969). These simplified equations provide for gradual increases in attenuation losses as a function of distance beyond the beginning of the shadow zone, and for a broad maximum in attenuation of 25 dB in the vicinity of 1,000 Hz, with lesser attenuation at either lower or higher frequencies.

The model considers downwind propagation, based on the experimental work of Franken and Bishop (1967). Excess attenuation caused by downwind propagation is, for practical purposes, limited to frequencies below 800 Hz.

Propagation Through Foliage and Vegetation

The literature (Eyring 1946, Wiener and Keast 1959, and Embleton 1963, Harrison 1974) shows conflicting results as to the effects of trees, brushes, and shrubs on the propagation of sound. However, recent work by Aylor (1972a,

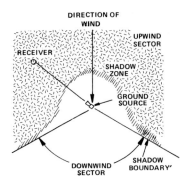

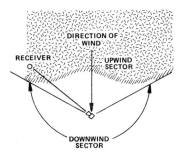

Figure 5b - Formation of sound zones resulting from typical
combination of (a) daytime, and (b) nighttime,
wind and temperature gradients.

1972b) provides a basis for the estimation used (Figure 6).
This relationship requires input data information concerning leaf area per unit volume, and leaf widths. The
program assumes, unless otherwise specified, a leaf area
per unit volume of 0.5 m^{-1}, and an average leaf width of
5 cm.

Shielding by Barriers

Figure 7 shows the model for barrier attenuation calculations. Such barriers provide a shadow zone; although
the shadow zone is not as deep as a similar shadow created
by a light ray, because the wavelength of sound is generally
of the order of magnitude of the dimension of the barrier,
and because of atmospheric scattering. The model selected
for the program is based largely upon the analysis of experimental data by Maekawa (1965, 1966). The approximation

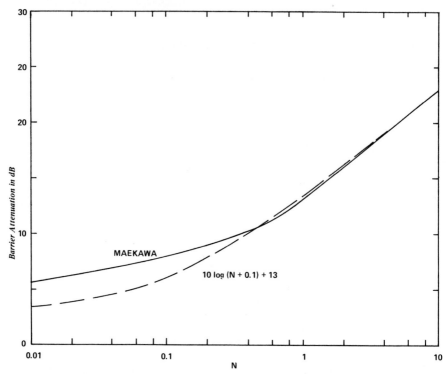

Figure 6 – Attenuation of noise by a rigid barrier.

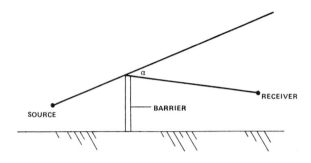

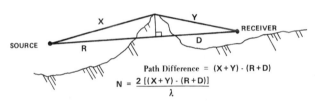

Figure 7 - Model for barrier attenuation calculations.

of Maekawa's data used by the program is shown in Figure 8.

DETECTABILITY AND REACTIONS TO AUDIBLE SOUNDS

      Presently, there is very limited knowledge of the threshold levels at which particular sounds are perceived by various wildlife species. Further, how these animals specifically react to particular sound is also little known. Thus, a project should be undertaken to determine at what level the noise created by certain human activities is detected by species of interest and the typical reactions by the species to the detected sounds. The parallel in the human animal to the concept that mere perception of certain sounds can give rise to a significant reaction can be found in the case of a mountain hiker who suddenly hears a motorcycle or snowmobile. Imagine the annoyance that is immediately felt. And, it is the information that the perception of the sound carries, rather than the sound itself, which gives rise to the reaction.

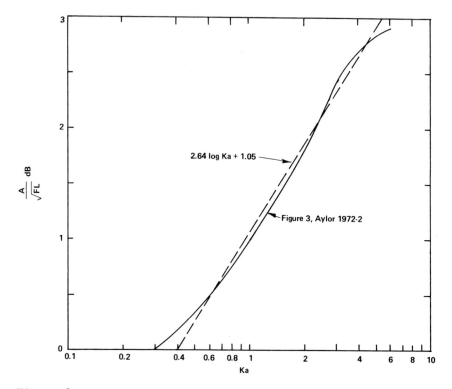

Figure 8 - Excess attenuation due to propagation through a
band of vegetation.

When studying wildlife detectability levels and reactions, one must also consider the acoustic impact on
animals in the food chain of the species of interest, as well
as animals peripheral to its survival. In summary, only if
the sound from a particular disturbance is determined to
be not audible, is further investigation of the extent of
the impact unnecessary.

Detectability Model

At first glance, it would appear that one could apply
the data developed by the engineering program previously
described to the threshold of hearing of a receiver animal,
and determine whether or not the animal detects the source.
This is an overly simplistic approach. The concept of
"threshold of hearing" is so firmly rooted in our academic
backgrounds that it seems almost sacreligious to suggest that

the concept does not apply, but I believe it does not.
With humans, at least, we know (from signal detection theory)
that the detectability of an acoustic signal with a noise
background is $d' = \mu \dfrac{s\omega^{\frac{1}{2}}}{N}$ where: μ is an expression of the efficiency of the observer with respect to an ideal energy
detector, s is the signal level in a one-third octave band,
N is the background noise level in the same one-third octave
band, and ω is the one-third octave bandwidth.

     In essence this tells us that detectability (or re-
ceiver "sensitivity") is a function of the observer's efficiency, the signal-to-noise ratio in a one-third octave
band, and the square root of the bandwidth of that band.
(For a formal treatment for the theory of detectability
see Swets 1964 and Green and Swets 1966).

     Implicit in this relationship is the view that de-
tection is a probabilistic, and not derministic, process.
Implicit also in this relationship is that detection decisions have two components--perceptual sensitivity, or
threshold of hearing, and response bias. Paraphrasing the
example given by Fidell and Bishop (1974) should make this
clear:

     The Audubon Club is holding a bird call listening
contest. The contest works like this. A group of contestants is placed in a forest glade and birds of various
species are stimulated to give forth their cries. The decisions facing our contestant are as follows:

	Actual Situation	
Contestant's Decision	Bird did not call	Bird called
Remain silent	Correct rejection	Miss
Identify bird	False alarm	Hit

     If our contestant remains silent when there is indeed
no bird, he has correctly rejected the hypothesis that a
bird has called. He has not given away any information and
preserves his chances to be a winner. If he remains silent
when a bird actually has called, someone else will identify
the bird and he will lose the contest. If he identifies
the bird when it was indeed not present, he has advertised
his expert knowledge to the use of other contestants.
Finally, if he identifies the bird when it first calls, he
will win the contest.

It is obvious that the costs and payoff associated with the four decision possibilities will have a major influence upon the apparent detectability of the bird call. It is also obvious that the probabilities of hit and false alarm will be affected by this same observer's bias.

Figure 9 shows a family of receiver operating characteristic curves.

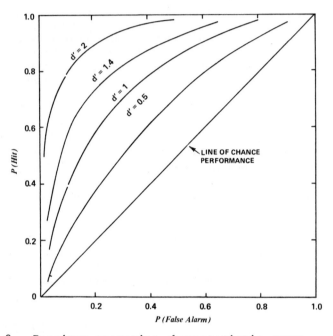

Figure 9 - Receiver operating characteristic curve.

Signal detectability theory assumes that for each parameter d' a unique set of probabilities of hit and probabilities of false alarm are associated. An observer can exhibit behavior corresponding to all points lying on the curve—from extremely poor, or conservative, detection performance (lower left-hand corner) to exceedingly good, or radical, detector performance (upper right-hand corner). Regardless of the values assigned to the observer's decision outcomes, all the detection behavior possible on one receiver operating characteristic curve can be shown to result from the same underlying threshold of physical hearing.

A practical difficulty arises in that the computer model used demands input of receiver sensitivity, d' (or

both probability of false alarm and probability of hit) and
observer efficiency. Because of human intellectual biases,
and the wide range of risks that detectability decisions en-
tail, these parameters are seldom known with any degree of
accuracy. As an illustration of the broad range of risks
which various observers will have to take into account when
making a signal detectability decision, consider two situations:

The first is a soldier listening for approaching combat
vehicles. If he incorrectly detects an approaching tank and
fires when none indeed approaches, he advertises his
position, thus making him extremely vulnerable to being
killed. This is a high-risk situation. Consider, next, the
environmentalist listening for off-road vehicle or chain
saw noise. Little is lost by incorrectly detecting the
noise, even when not present. This is a low-risk situation.

Applicability to Animals

The above trivial example becomes meaningful if we
consider caribou grazing on winter range. The sound source
is that of a stalking wolf. If the wolf is not present,
the caribou hears nothing and decides not to bolt. He
has conserved valuable energy and made a correct decision.
If the wolf is indeed present, but the caribou decides a-
gainst running, he will be devoured. If the caribou bolts
each time he thinks he hears a wolf, whether one is there
or not, he will soon run off all his fat and expire any-
way. Finally, if the wolf is indeed present and the
caribou bolts, he may be a bit skinnier--but at least he
survives another day.

The applicability of signal detectability theory to
animals has not been substantiated. However, one factor
simplifies the applicability as compared to the use of signal
detectability theory for human observers. With human observers, as illustrated by the examples given above, detector
efficiencies and probabilities of hit and false alarm are
seldom known and vary greatly. With animals, hopefully,
these probabilities will be built in by instinct, or at
least habituated in some predictable manner.

The program makes the prediction of detectability
based upon the relationship: $10 \log (S/N) = 10 \log \dfrac{d}{N \, \omega^{\frac{1}{2}}}$.

At each interrated distance, the program determines whether
the conditions of this equation have been met or not. If
the left side of the equation is greater than the right,
then the sound is detectable. If the right side is greater,
then it is not. When the condition is met, then the sound
is just detectable. Nothing in the equation prevents the
program from predicting detectability at signal levels far
below the animal's threshold of hearing. Thus the program
contains a device which limits the detectability distance
to the absolute threshold of hearing of the hearer animal.

The program leaves unanswered the important question:
"What are the effects of intrusive sound, above the limit
of detectability, on wildlife?" It only proposes to say
whether or not the animal hears the sound, thus setting an
"outer limit" of noise impact on species of concern.

Reynolds (Reynolds 1974, McCourt, et al 1974) observed that Dall sheep seemed unaffected at locations where
intrusive noise was faintly audible to humans. The sheep
from the same group showed some reaction, however, at
locations where the noise levels were higher (and thus
presumably detectable). Did the sheep actually hear the
sounds at the more remote location? The detectability
approach attempts to answer that question.

Of course, it is entirely possible that the concept of
d' detector "sensitivity", is meaningless for animals. A
deterministic approach to detectability could then be proper.
The program is presently being modified to provide a print-
out of contours of equal S/N, and $S/N\omega^{\frac{1}{2}}$, (for the one-
third octave band with the highest S/N at the hearer's
location). If masking in animals is similar to masking in
humans, and if one-third octave bands are similar to critical
bands in animals, as they are in humans, then this approach
will provide a graphic measurement of the extreme limits
on animals of the acoustic impact of intrusive activites.

SUMMARY

The engineering prediction portion of the program has
been fairly well validated, as has the detectability part
using human observers. However, as previously mentioned, no
attempt has yet been made to validate the theory of signal
detectability as applicable to animals. To validate or
disprove this theory, perhaps the most direct approach would
be to first determine an acoustic stimulus known to give rise
to a violent reaction in the test animal. The signal-to-
noise ratio at the hearer animal's ears would be known.

From observations of reactions, probabilities of hits and
false alarms could be determined, thus defining d$'$andμ, for
at least this combination of circumstances.

Next, the hearer animal in his habitat should be subjected to the noise of the activity the impact of which is
of concern. If reaction is noted at detectability levels,
mere detectability is a wildlife impact, and further evaluation of that impact will be needed. If the level at the
animal's ear must be increased before reaction is noted, then
the "engineering" program can predict the levels needed to
produce reaction.

Using this program, we are able to predict with some
accuracy spectral and temporal patterns of sound at distance
remote from the source, and to predict their impact on
humans. Higher animals, like snakes, as Mark Twain (1962)
said, present a different problem. What do they hear? Or,
said differently,what is the acoustic impact of our activities on them? I feel that the detectability program will
provide a more accurate measure of the outer limit of this
impact than a deterministic threshold approach.

LITERATURE CITED

1. Aylor, D. 1972a. Noise reduction by vegetation and ground. J. Acoust. Soc. Am. 51:197-205.
2. Aylor. D. 1972b. Sound transmission through vegetation in relation to leaf area density, leaf width and breadth of canopy. J. Acoust. Soc. Am. 51:411-414.
3. Delany, M.E. 1969. Range predictions for siren sources. Aero. Spcl. Rep. 033. Natl. Phys. Lab.
4. Embleton, T.F.W. 1963. Sound propagation in homogeneous deciduous and evergreen woods. J. Acoust. Soc. Am. 35(8):1119-1125.
5. Eyring, C.F. 1946. Jungle acoustics. J. Acoust. Soc. Am. 18(2):257-270.
6. Fidell, Sandford and Dwight, E. Bishop. 1974. Prediction of acoustic detectability. TARADCOM Tech. Rep. 11949. U.S. Army Tank-Automot. R&D Comm., Warren, Mich. 48090.
7. Franken, P.A. and D.E. Bishop. 1967. The propagation of sound from airport ground operations. NASA CR-767.
8. Green, D.M. and J.A. Swets. 1966. Signal detection theory and psychophysics. John Wiley & Sons, New York, NY. 10016.
9. Harrison, Robin T. 1974. Sound propagation and annoyance under forest conditions. ED&T Rep. 7120-6. USDA For. Serv. Eqpt. Dev. Ctr., San Dimas, Calif. 91773.
10. Harrison, Robin T. 1975. Impact of off-road vehicle noise on a National Forest. Proj. Rec. ED&T 2428. USDA For. Serv. Eqpt. Dev. Ctr., San Dimas, Calif. 91773.
11. Horonjeff, Richard and Sanford Fidell. 1977. Prediction of acoustic detection ranges for multiple sources and spatially distributed detectors. TARADCOM Tech. Rep. 12240. U.S. Army Tank-Automot. R&D Comm., Warren, Mich. 48090.
12. Ingard, U. 1953. A review of the influence of meteoroligical conditions on sound propagation. J. Acoust. Soc. Am. 25:405-411.
13. Ingard, U. 1969. On sound-transmission anomalies in the atmosphere. J. Acoust. Soc. Am. 45:1038-1039.
14. Ingard, U. and G.C. Maling, Jr. 1963. On the effect of atmospheric turbulence on sound propagated over ground. J. Acoust. Soc. Am. 35:1056-1058.
15. Kneser, H.O. 1933. Interpretation of the anomalous sound absorption in air and oxygen in terms of molecular collisions. J. Acoust. Soc. Am. 5:112-126.
16. Maekawa, Z. 1965. Noise reduction by screens. Memoirs of Facl. of Engr., Kobe Univ., Japan 11:29-53.
17. Maekawa, Z. 1966. Noise reduction by screens of finite size. Memoirs of Facl. of Engr., Kobe Univ., Japan 12:1-12.

18. McCourt, K.H., J.D. Feist, D. Doll, and H.J. Russel. 1974. Disturbance studies of caribou and other mammals in the Yukon and Alaska, 1972. Can. Arctic Gas Study Ltd. & Alas. Arctic Gas Study Co.

19. Piercy, J.E. 1969. Role of vibrational relaxation of nitrogen in the absorption of sound in air. J. Acoust. Soc. Am. 46:602-604.

20. Reynolds, Patricia C. 1974. The effects of simulated compressor stations sound on Dall sheep using mineral licks on the Brooks Range, Alaska. Biol. Rep. Ser. 23, Can. Arctic Gas Study Ltd. & Alas. Arctic Gas Study Co.

21. Rudnick, Isadore. 1957. Propagation of sound in the open air. In Handbook of noise control. Cyril M. Harris, ed. McGraw-Hill Book Co., New York, NY 10020.

22. Society of Automotive Engineers. 1964. Standard values of atmospheric absorption as a function of temperature and humidity for use in evaluating aircraft flyover noise. Aerosp. Recom. Pract. 868. Soc. of Automot. Engr., Warrendale, Pa. 15096.

23. Society of Automotive Engineers. 1974. Acoustic effects produced by a reflecting plane (draft) Aerosp. Info. Rep. 1327. Soc. of Automot. Engr., Warrendale, Pa. 15096

24. Sutherland, L.C. 1973. Sound propagation over open terrain from a source near the ground. J. Acoust. Soc. Am. 53:339.

25. Swets, J.A.1964. Signal detection and recognition by human observers. John Wiley & Sons, New York, NY 10016.

26. Twain, Mark. 1962. Letters from the Earth. Bernard deVoto, ed. Harper & Row, New York, NY 10022.

27. U.S. Environmental Protection Agency. 1971. Effects of noise on wildlife and other animals. Rep. NTID300.5 USEPA Off. of Noise Abate. & Contr., Washington, DC 20460

28. Wiener, F.M. and D.N. Keast. 1959. Experimental study of the propagation of sound over ground. J. Acoust. Soc. Am. 31:724-733.

NOISE AND ANIMALS:
PERSPECTIVES OF GOVERNMENT AND PUBLIC POLICY

Raelyn Janssen

Office of Noise Abatement and Control
U.S. Environmental Protection Agency
Washington, D.C.

INTRODUCTION

Life scientists traditionally undertake research with the goal of learning about mechanisms of normal behavior or physiological function. That is, the question often asked is how, or by what means, does a species maintain normal life patterns in the natural environment?

Policy-makers, on the other hand, most often need to know how much, i.e. given different amounts of a stimulus, how much effect is observed? Further, the emphasis is not on the natural condition, but is directed toward comparing the natural or pre-existing condition with those that will result from some human modification of the natural state. This information must then be translated into a form suitable for comparing the costs and benefits of a set of alternative project actions. Thus, for the decision-maker concerned with noise, the primary requirement is quantification of the benefits (to wildlife or humans) of different noise-producing activities or control alternatives.

I. EVALUATION OF IMPACT

Information necessary to quantify impact of noise or of noise control on the animal kingdom has not been developed yet. Work on impact-modelling over the last few years has produced models capable of quantifying some of the impact of noise on humans. Corresponding work on wildlife impact will undoubtedly require relatively more research for a number of reasons, including:

287

- Human communications and auditory function are very well understood by comparison with those of other species.

- The vast diversity of the animal kingdom makes it impossible to speak in general terms of noise effects on animals. The thousands of species of birds, insects, mammals, and others vary greatly in physiology, habitat, and behavior patterns. In consequence, they also vary in the types of environmental noise sources which affect them.

At first glance, complexities produced by such variety may appear to preclude an orderly account of noise effects. However, if animals are grouped according to similarities of behavior and habitat, it is possible (1) to establish types of noise sources which may impinge on each of these groups, and (2) to suggest possible impacts of legislation and regulation. Figure 1 gives a suggested framework for the study of effects of noise on wildlife, which may account for diveristy between species and geographic differences in noise control.

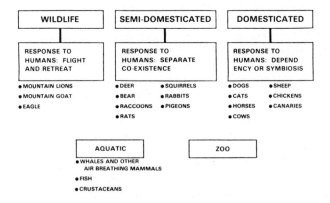

Figure 1 - Animal response to humans by category.

The framework is founded on the basic reaction of an animal to human activity.

The model has five categories. The first three categories are wildlife, semi-domesticated animals, and domesticated animals. The last two categories are special cases: animals in zoos and aquatic animals.

A. Wildlife

For purposes of evaluating noise effects, wild animals are defined as those that tend to flee from human intrusion. These animals are assumed to be exposed to relatively few unnatural noises and little long-term noise. Most of their exposures are involuntary and come from mobile sources such as snowmobiles, bush planes and motorcycles. It is possible that response parameters such as probability and magnitude of startle, or likelihood of interruption of normal feeding, nesting or reproduction may distinguish this animal group. Fixed sources such as railroads, pipelines, and transmission lines tend to force wild animals back into quieter territory. The greateast effect of noise on this group may therefore be indirect: reduction of land area available for habitation and concomitant reduction of population size.

B. Semi-Domesticated

The semi-domesticated category refers to animals which have voluntarily permitted the human technological environment to envelop them, enduring additional dangers and problems in exchange for benefits such as food. For some urban semi-domesticated animals such as squirrels and raccoons, the noise environment is similar to that experienced by urban and suburban human inhabitants. Like humans, these animals receive considerable noise from streets and highways, airports, and construction sites. Some of these animals appear to adapt to these noises, or to the noise sources themselves, but, as is true for humans, the costs of adaptation remain undefined.

It is interesting that for this class, noise is sometimes used as a noxious deterrent because the animal is considered a pest. An example is the use of gas-fueled explosive devices to frighten away birds from orchards and airport runaways.

C. Domesticated

The category of domestic animals includes pets and farm animals. The noise exposure of this group is usually involuntary. Examples are farm animals or pets which are confined to noisy areas. Noise exposures of domesticated animals range from typical urban noise patterns to specialized patterns such as those produced by farm equipment.

D. Aquatic

The aquatic medium presents a different set of acoustic and physiological conditions which require study. For example, fish perceive sound via the lateral line, exposed on the suface of the body; at this time we do not know the susceptibility of this organ to damage by noise. Prominent noises for the aquatic category include waterbone sound from boats and ships (potentially affecting all species) and possibly sonic booms, which may affect species that spend time near the surface of the sea, such as aquatic mammals.

E. Zoo

The zoo is a special case where wild species and others involuntarily reside in a domestic setting. This results in modifications of natural behavior patterns. It may be assumed that frequently the zoo environment is noisier than the native habitat. Here it may be very important to distinguish between animals that were caught wild and have now been placed in the zoo, and those animals that were born in captivity. Keeping this distinction in mind, the zoo represents an important laboratory for further research on certain species.

IMPACT OF EXISTING NOISE-RELATED POLICY ON ANIMALS

It may be conjectured that existing noise regulation around the world has already had some impact on the noise climates experienced by various animals. Although no regulations or laws in any country have been passed solely with the purpose of protecting animals from noise (EPA, 1976), many of the regulations may have indirectly improved or degraded the noise environments of animals.

It should be noted that, for the most part, only the developed nations have imposed restrictions on noise emission. This is fitting in that these same countries are largely responsible for the world-wide proliferation of noise sources.

But control of noise, which has been slow in developing, has focused on protecting human populations within the developed countries. Consequently, regulations of noise emission are protective of wildlife only where the interests of human and animal populations coincide. This is illustrated in Figure 2, which indicates likely impacts of noise regulatory activities on each category of animals.

TYPE OF LAW	WILDLIFE	SEMI-DOMESTIC	DOMESTIC	AQUATIC	ZOO
SONIC BOOM	YES (ADVERSE)	YES (POSITIVE)	YES (POSITIVE)	YES (ADVERSE)	YES POSITIVE
MOTOR VEHICLE REGULATIONS	YES	YES	YES	NO	YES
CONSTRUCTION AND LOGGING NOISE	YES	YES (EUROPE, USA, JAPAN)	YES	NO	YES
AIRPORT REGULATIONS AND LAWS	RARELY	YES	YES	NO	YES
BIRD DEVICES	NO	YES (ADVERSE)	YES (ADVERSE)	NO	NO
COMMUNITY NOISE REGS (NOISE ZONING)	NO	YES	YES	NO	YES
LAND USE PLANNING	YES	YES	YES	NO	YES
BUILDING REGULATIONS	NO	NO	YES	NO	NO
SNOWMOBILES AND OFF-ROAD VEHICLES	YES	YES	YES	NO	NO
AIRCRAFT REGULATIONS (NON-SST)	YES	YES	YES	NO	YES
POWER BOAT REGULATIONS	NO	NO	NO	ALMOST NONE	NO

Figure 2 - Existing restrictions which may influence noise
     environments of animals.

A. Sonic Boom

    Current sonic boom controls in some countries including
the United States and several Western European countries have
effectively banned commercial supersonic flights over in-
habited land areas (EPA 1976). Military supersonic flight is
also confined to less populated areas in many countries. As
a result, sonic booms affect land and marine animals which
ordinarily have little exposure to manmade noise.

B. Motor Vehicle and Highway Noise

    General limitation of highway noise is the most wide-
spread form of noise regulation (EPA 1976). However, there
is now a trend toward setting maximum noise emissions for
new vehicles. In 1957, the Economic Commission for Europe
(ECE) published a model regulation with noise limits for
different types of vehicles, which was revised in 1974. Most
of the ECE states, as well as Japan, Yugoslavia and Czechos-
lovakia, have adopted these noise limits. In addition, the
Common Market countries have issued a schedule of nearly
identical noise limits for different types of vehicles. In
North America, the United States has promulgated a noise regulation affecting new medium and heavy trucks, with limits
on buses, motorcycles and light vehicles to follow. These
regulations appear to benefit all except aquatic animals.

C. Construction Noise

    The United States, France, Germany, Switzerland and
Austria, among others, have issued regulations limiting the
noise from new construction equipment. The most widely
regulated construction noise source is portable air compres-
sors. Some countries also regulate less stringently existing
construction equipment of various types. Switzerland, for
example, has a permit system, under which operators of
noisier equipment just barely meeting the limits may be
denied permits to work on certain sites. This could have the
effect of driving the noisier equipment into rural and semi-
rural areas, increasing the noise exposure of animals.

D. Airport and Aircraft Regulations

    Most of the impact of aircraft noise occurs near air-
ports, although impact away from airports also takes place,
particularly in the case of lowflying aircraft such as crop-
dusting planes and bush planes. Several countries have strict
regulations on use of such planes. One of these is Switzerland, where only quiet types of propeller planes may be

authorized to operate in certain parts of the country. The
United States has proposed tighter noise limits for new small
propeller planes which, if adopted, would eventually have an
effect on planes being used as bush planes. Such regulations
are anticipated to have a beneficial effect on all types of
animals.

E.   Community Noise Regulations

Community noise regulations generally are intended to
control noise from different types of sources as a function
of land use: residential, industrial, commercial, hospital,
etc. Some governments instituted guidelines to be used by
towns and cities in the development of community noise regulations. For example, in Denmark, Switzerland, the United
Kingdom and Germany, there are recommended limits for recreation areas and parks, which are classified with the most
noise-sensitive zones. Typical guideline values for ambient
noise levels are energy-equivalent levels of 45 dBA in daytime and 35 dBA at night in these areas. To the extent that
the guidelines are effective, the noise exposure of animals
living in these zones is reduced.

F.   Land Use Planning

In some countries, such as Japan, the United States,
France, and the United Kingdom, there are now restrictions
on building housing too close to transportation corridors,
particularly highways, or, conversely, on construction of
new highways close to existing housing. In the U.S., all
new federally-funded highways must meet noise standards at
the right-of-way boundary. Where space permits, a buffer
zone may meet this requirement. In the USSR, similar zones
are required for new factories if they emit noise or noxious
gases. The buffer zones may be in the form of commercial
space for non-sensitive uses such as warehousing or they may
be in the form of greenbelts. If good wildlife cover is
created in buffer zones adjacent to noise sources, then the
noise exposure of some opportunistic semi-domestic animals
may actually increase as they move in to take advantage of
the new habitat.

Of course, there are many different activities which
constitute land-use planning, and an almost limitless range
of potential animal impacts.

G.   Snowmobiles and Other Off-Road Vehicles

Regulation of OffRoad Vehicles (ORVs) has probably had
the greatest impact in reducing noise exposure of animals in
certain countries. Most snowmobiles now in use are in Canada
and the United States. Within the last few years, snow-
mobile noise has been regulated by State governments in the
U.S. and in Canadian provinces. Before these regulations,
noise levels from accelerating snowmobiles were typically
up to 100-102 dBA measured at fifty feet. At present,
most new snowmobiles can comply with the typical U.S. and
Canadian limits of 78 dBA at fifty feet.

Another wilderness noise problem is the off-road motorcycle, or "trail bike". In Europe there are no special regulations for these vehicles, and the emission limits applicable to road motorcycles also apply to those being used off
of roads. However, off-road use in Europe is less widespread
than in North America and is often forbidden outright. In
West Germany, for example, if a motorcycle is found being
used off-road, it is subject to confiscation. In the United
States, Canada, and Australia, however, specially designed
off-road motorcycles may measure 5 to 10 dBA more than their
"street-legal" cousins. At present, two U.S. States have
regulations on off-road motorcycles: California and Oregon.
In Oregon, off-road cycles must be as quiet as street-legal
cycles, while in California, the level may be 3 dB higher.

Besides noise limits, there are also outright bans
on trail bikes on some public land. In the United States
two different Departments control such use: Department
of Interior for the national parks and general public lands,
and Department of Agriculture for national forests. It is
estimated that Interior's Bureau of Land Management controls
twenty percent of all U.S. land area, and over one half of
the land which is being intensively used by ORVs. That
agency is now developing policy to govern decisions about
ORV use. Recently, President Carter showed interest in the
problem by revising an Executive Order to clarify agency
authority to define zones of use by off-road vehicles on
public lands (The White House, 1977). Among other things,
this order requires agencies to ban ORVs immediately whenever it is determined that considerable adverse effects are
being caused by them. Such adverse effects are defined in
the Executive Order to include effects on wildlife or the
wildlife habitat.

H. Power Boats

Noise from motorboats is virtually unregulated. Exceptions are Austria's operating limits of 70 dBA at 25 meters, the State of New Jersey's new standard, and some lake curfews in West Germany.

GOVERNMENT AND PUBLIC POLICY NEEDS

Noise produced by transportation, recreation and construction sources is modifying the environments of animal species. It is assumed that these changes in auditory environment affect wildlife. But in order to inform public policy we must know both the extent of the effects of noise, and the meaning of that effect in measurable terms. For animals as well as for the human public, government decision makers need noise criteria--that is, quantified relationships between the amount and pattern of exposure to noise, and the measurable effect on the subjects.

Figure 3 shows two examples of criteria which are used to assess human effects of noise (EPA, 1974).

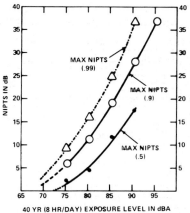

NOISE-INDUCED HEARING LOSS (NIPTS) AS A
FUNCTION EXPOSURE LEVEL FOR THE
50TH, 90TH AND 99TH PERCENTILES

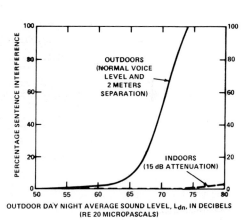

MAXIMUM PERCENTAGE INTERFERENCE WITH
SENTENCES AS A FUNCTION OF THE
DAY-NIGHT AVERAGE SOUND LEVEL

Figure 3 - Examples of noise criteria for human effects.

The top figure shows hearing loss, or Noise-induced Permanent Threshold Shift (NIPTS), for different exposures. The bottom figure shows anticipated interference with speech communications across exposure values. Criteria such as these are used in comparing and justifying noise control activities, weighing "benefits" of noise control against the costs. In quantifying noise effects on animals, however, specification of both dose and response raise considerable problems.

A. Describing Exposure

For human criteria, exposures are typically described in terms of both cumulative measures and measures of detectability or intrusiveness. An energy-averaged equivalent sound level (L_{eq}) is a commonly used cumulative descriptor. In the U.S., the day-night average sound level, L_{dn}, is also in frequent use. L_{dn} is a variant of L_{eq} which incorporates a 10-dB nighttime penalty to more heavily weight sleep-distrubing noise. It is also necessary to describe single events of noise in terms which describe its intrusion into the acoustic environment. Many descriptors are available for this purpose.

A critical dimension of any descriptor is its frequency-weighting. L_{eq} and L_{dn}, as used, incorporate A-weighting, which works well for predicting human effects of noise, largely because it was derived from human equal-loudness curves. Clearly, use of the A-weighting as a descriptor of impact on the animal kingdom must be questioned. The variety of frequency sensitivities and auditory ranges is so vast that one weighting system applicable to, for example, insects, bats and moose may be impracticable.

Beyond the choice of descriptors, there are questions about quantifying exposures. There are two alternatives: direct measurement of noise at or near the ear, and calculated exposures based on models of sound propagation in the animal's environment.

Direct measurement is, of course, the preferable alternative, but it poses problems of instrumentation. It may be possible to outfit some larger species with the small tape recorders and dosimeters available today if the equipment can be adequately secured. But for the most part, it would be necessary to develop specialized instrumentation for animals. Calculation or prediction of exposure is also difficult, especially for animals which run free.

A basic element of characterizing noise exposure is describing habitat background noise. This is necessary to determine the degree of intrusion of a noise above the natural background, and the masking produced by the noise. Figure 4 shows the spectra of different noise sources in a particular wildlife habitat in Australia.

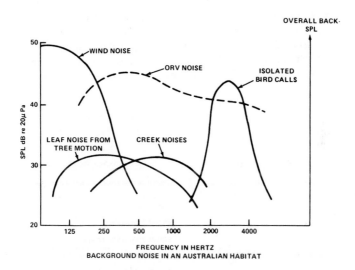

BACKGROUND NOISE IN AN AUSTRALIAN HABITAT

Figure 4 - Maximum percentage interference with sentences as a function of the day-night average sound level.

Off-road vehicle noise may be compared with natural sounds, as determined by Bennison and Wallace (1976).

B.   Describing the Response

     Considerations on the response side of the dose-response relationship are even more difficult. Irrespective of the organism, there are two questions: What exactly does the noise do, and what is the functional importance or meaning of it? It may be helpful here to introduce the concept of primary effects and secondary effects. Primary effects are the immediate effects of the noise on the organism. The secondary effects are the whole range of consequences of a primary effect. In Figure 5 are examples of some primary and secondary effects of noise on people.

PRIMARY EFFECTS	SECONDARY EFFECTS
MASKING OF COMMUNICATIONS	LOSSES IN THE EDUCATIONAL PROCESS
	ACCIDENTS DUE TO MISSED WARNING SIGNALS
	SOCIAL ISOLATION
	STRESS AND ANNOYANCE
NOISE-INDUCED HEARING LOSS	ISOLATION FROM FAMILY AND FRIENDS
	REDUCED EMPLOYABILITY
	DISCOMFORT AND PAIN FROM MODERATE-LEVEL NOISES
	ACCIDENTS DUE TO MISSED WARNING SIGNALS

Figure 5 - Sample effects of noise on people.

For the various species of animals, there could be a multitude of such effects. Primary effects, masking and so forth, may be essentially the same across species, including homo sapiens. Secondary effects, however, will be very different for non-human species because they may be interruptions in important life functions, such as nesting, migration, and hibernation.

Figure 6 shows a few hypothetical primary and secondary effects of noise in some animals.

SPECIES	PRIMARY EFFECT	SECONDARY EFFECT
BIRDS	MASKING OF SIGNALS	INTERFERENCE WITH MATING?
SMALL ANIMALS, SEMI-DOMESTIC	MASKING	EFFECT OF KILL RATIOS IN SPECIFIC PREDATOR-PREY PAIRS? (DEPENDS ON WHICH RELIES MORE ON HEARING AS OPPOSED TO SIGHT AND SCENT)
DOMESTIC AND SEMI-DOMESTIC	HEARING LOSS	SAFETY, MATING?
AGRICULTURAL, DOMESTIC	STRESS RESPONSE	CHANGES IN EGG-LAYING, MILK PRODUCTION, WEIGHT-GAIN?

Figure 6 - Some possible primary and secondary effects in
           some animals.

Clearly, both types of response information are necessary to evaluate the impact of noise on animal species. The primary effects are the means by which disruptions of life functions can occur. These effects are often best studied in the laboratory. On the other hand, quantifying the disruption of life activities, or secondary effects, requires work in the field. This complicates the instrumentation problem posed above, because, in addition to tape recorders or other exposure recording devices, animals may need to wear telemetry devices, or other apparatus for remote sensing of responses. The researcher will need to develop techniques for unobtrusive observation.

Regardless of measurement techniques, the investigator will need to consider these questions:

- Does adaptation or habituation take place, and if so, what form does it take?

- Can normal behavior be determined without interaction with the techniques of observation?

- Is "modified behavior" ever solely attributable to noise? For example, does an animal run from the noise itself or from the presence of people or machines that is merely signalled by the noise?

- When modified behavior due to man-made environmental noise is established, is there an adverse effect? For example, if a migratory pattern is changed, is that change detrimental?

CONCLUSION

Human activities are changing the acoustic environment world-wide, potentially affecting all animal species. It is important that we understand the impact of noise so that decisions we make now can avert detrimental effects on the animal kingdom. But, for practical purposes, understanding impact requires knowing both (1) how an effect occurs, and (2) how much effect occurs with a given dose. In answering basic research questions about life mechanisms, it is often possible to contribute to the bank of quantified data which tell us how much. Specifying both dose and response contributes to the decision-maker's capability to make good policy.

REFERENCES

1. Bennison, David C. and Wallace, Allan K. "The Extent of Acoustic Influence of Off-Road Vehicles in Wilderness Areas". Paper presented at the National Symposium for Off-Road Vehicles in Australia, Canberra, January-February, 1976. Adelaide, S.A.: Department of Mechanical Engineering, University of Adelaide, p. 4.

2. EPA. Information on Levels of Environmental Noise Requisite to Protect Public Health and Welfare With An Adequate Margin of Safety. Superintendent of Documents, U.S. Government Printing Office, Washington,D.C. 20402 (1974).

3. EPA. 1976 Reassessment of Noise Concerns of Other Nations, EPA 550/9-76-011, -012, August, 1976.

4. The White House, Executive Order No. 11644, as amended May 24, 1977.

SUMMARY AND DISCUSSION

Upon completion of presentation of the papers at the symposium, a workshop-round table discussion of the issues and problems associated with noise and wildlife was convened. This workshop round table discussion was most useful. It allowed members of the audience attending to contribute and, in a sense, summarized and integrated much that had been presented during the two days of the symposium.

One of the issues raised and addressed was that of coordinating the efforts of the scientist who does research aimed at determining the various effects of noise on wildlife with the various agencies charged with protecting wildlife. Certain scientific coordinating bodies exist, such as the National Academy of Sciences, National Research Council Committee on Hearing, Bio-Acoustics and Bio-Mechanics (CHABA) in the United States but that group is not specifically charged with responsibility for wildlife and would only assume a mission in this area if requested by one of its sponsors.

The experience of EPA has clearly shown the problems involved in going from off the shelf scientific results to formulation of damage risk criteria, then to regulations governing exposure to various potentially hazardous agents. During the course of discussion, the point was made that regulations rather than legislation might be more effective in controlling noise.

It became clear during the course of the symposium that North America was more concerned about noise and wildlife than any other area. This is not hard to understand when you realize that most of the wildlife in developed countries is found in Canada and the United States. In western Europe, most of the wildlife are primarily concentrated in mountainous areas, national parks and such places. Also, Western Europe is densely populated, much more so than most of the U.S. and Canada.

303

Another point that emerged during discussion was that decibels on the A scale (dBA) was not the most appropriate measure to use in dealing with noise levels and their effect upon wildlife. That measure has been most successfully applied to various effects of noise on man, such as hearing loss, annoyance, etc. but it is not appropriate for wildlife for many reasons. It was recommended that sound pressure level (SPL) be used until research pinpoints the most appropriate measure or measures.

There was some consensus among those present on several aspects of the problem. With regard to research needs, it was generally agreed that there is a critical need for more scientific data. There is not overwhelming evidence that noise is not a significant adverse factor in the life of animals. But neither is there clear definitive data demonstrating that noise is an adverse stimulus or stressor.

Research needed to answer critical questions in this area would include:

1. Study of individual species, one by one, not only as individual animals, but in social groups (herds, flocks, etc.) such studies should examine the acoustic nature (frequency, intensity, temporal pattern etc.) of critical events of the animal (mating, territoriality, alarm, nurture, etc.).

2. More complete knowledge of the spectrum of environmental sound is necessary. Presently, most noise analyses cover only a limited frequency range and do not include areas that could be critical to many animals.

3. We also need to know the effect of noise upon a given animal population that is declinig, regardless of why the population is declining.

4. The combined stressor effects of noise with other stresses upon an animal should be studied because of possible potentiation.

5. Both long and short term noise effects must be studied.

6.   Studies should include both field and laboratory ef-
      forts. Each has merit and neither is sufficient by
      itself, with each type providing feedback and useful
      information to the other.

7.   Further studies of sound propagation in the field
      must be done, because this could also be critical.